The Chemical Scythe
Lessons of 2, 4, 5-T and Dioxin

DISASTER RESEARCH IN PRACTICE
Series Editor: Frances D'Souza
International Disaster Institute
London, United Kingdom

THE CHEMICAL SCYTHE: Lessons of 2,4,5-T and Dioxin
Alastair Hay

The Chemical Scythe
Lessons of 2,4,5-T and Dioxin

Alastair Hay
The University of Leeds
Department of Chemical Pathology
Leeds, England

PLENUM PRESS • NEW YORK AND LONDON

Library of Congress Cataloging in Publication Data

Hay, Alastair, 1947—
The chemical scythe.

(Disaster research in practice)
Bibliography: p.
Includes index.
1. Chemical industries—Accidents. 2. Chlorophenols—Toxicology. 3. Tetrachlorodibenzodioxin—Toxicology. 4. Chlorophenols—Environmental aspects. 5. Tetrachlorodibenzodioxin—Environmental aspects. 6. Air pollution. I. Title. II. Series.
RA578.C5H39 1982   363.1'79   82-12249
ISBN 0-306-40973-9

© 1982 Plenum Press, New York
A Division of Plenum Publishing Corporation
233 Spring Street, New York, N.Y. 10013

All rights reserved

No part of this book may be reproduced, stored in a retrieval system, or transmitted in any form or by any means, electronic, mechanical, photocopying, microfilming, recording, or otherwise, without written permission from the Publisher

Printed in the United States of America

To Wendy and Tom

Foreword

The Chemical Scythe is the first book in a projected series to be published by Plenum Press in association with the International Disaster Institute. The aim of the series, *Disaster Research in Practice*, is to provide scientific and readable accounts on the most urgent areas of disaster research. It is fitting, therefore, that Dr. Hay's investigation into the nature and effects of dioxins heralds the new series.

The problem of chemical hazards is one that we will have to learn to live with in future decades. Dr. Hay's book is an authoritative account of the chemistry and proven and potential effects of dioxins, and of the implications for safety planning. He concludes with a cautious, yet optimistic note—that indeed we can learn to live with such hazards, providing that we are prepared to understand and plan for the unexpected.

The accident at Seveso in 1976 alerted the world to an imperfectly understood but immensely alarming environmental hazard. Public debate and argument as to the implications of dioxins and, indeed, the use of herbicides as aggressive weapons in Vietnam, rage on. And yet it is only through the painstaking research exemplified in this book that it will eventually be possible to promote the vital accountability on the part of industrialists and governments.

The blow-by-blow story of Seveso is, in itself, gripping enough to allow the reader to forget that he is reading science. Equally, the example of Love Canal—a well-documented case of what not to do—should prompt all citizens at risk from chemical hazard to become sufficiently informed and thereby ensure greater safety for themselves and their families.

This book, in that it addresses the chemical nature of such hazards, as well as the health and social effects, provides a sound basis for future health and safety plans. It will be of immediate value to industrialists, manufacturers, government departments, administrators, planners, and, of course, academic scientists throughout the world. The content matter and the way in which it is told make the whole story accessible to the ordinary reader. Although there have been a few scientific accounts of the Seveso accident and its aftermath, this is the first book to take a profound look at chemical hazard, the use of herbicides, the question of public accountability on the part of large companies, and the increasingly urgent problem of toxic waste.

Dr. Hay's book is very much in the mold of the research prompted by the International Disaster Institute, the philosophy of which can be simply stated: A hazard cannot be dealt with in any sense of the word until its dimensions are properly known. It is our intention to continue our inquiries into disasters, their antecedents and their consequences, in the hope that the decision makers will become better informed and the public better able to protect themselves.

Frances D'Souza
International Disaster Institute
November 1981

Acknowledgments

This book was made possible by the generous assistance of numerous people. Professor Egbert Pfeiffer of the University of Montana and Dr. Guiseppi Reggiani of F. Hoffman-la Roche have provided me with invaluable information for several chapters in the book.

My involvement with the venture began with a series of articles in the scientific journal *Nature*. The editorial staff of the journal have given me every encouragement, and I am indebted to Dr. David Davies (the previous editor of *Nature*), to Chris Sherwell (former news editor), and to Judy Redfearn.

The staff in the library of *The Star* in Sheffield made their clippings file available to me for background information on one of Britain's dioxin accidents. Without this information the chapter on Coalite and Chemical Products Ltd. would have been much briefer and less detailed.

Professor Arthur Westing of Hampshire College, Massachusetts reviewed the chapter on herbicide spraying in Cambodia, and also supplied a unique photograph for the book. I am indebted to him for this help, just as I am to Terry Collins of the Graphics Department of the Polytechnic of Leeds who is responsible for most of the artwork in the book.

Dr. Frances D'Souza, Director of the International Disaster Institute, has my grateful thanks for her commitment to this project and for making the book possible in the first place. Lynn Ruffell of IDI transformed my various handwritten drafts into the final manuscript and she, too, has my grateful thanks.

One's editor is expected to assist with the preparation of the manuscript, but of course editorial involvement goes far beyond this. Criticizing a manuscript requires the finesse of a diplomatist: fortunately, my editor Ken Derham has this gift! His recommendations for changes to the original manuscript have given more substance to the book, and I am thankful that I have had his advice.

Wendy, my wife, made comments on, and edited my manuscript. It reads all the better for her involvement. The index is also her work, and I cannot thank her enough.

Alastair Hay

Contents

Introduction		1
1.	Chemistry and Occurrence of Dioxins	5
2.	Toxicology of Dioxins	25
3.	Hexachlorophene	69
4.	2,4,5-T, Trichlorophenol, Pentachlorophenol, and Polychlorinated Dibenzofurans	77
5.	Chloracne	89
6.	Industrial Accidents in Trichlorophenol Manufacture	95
7.	Vietnam and 2,4,5-T	147
8.	2,4,5-T in Cambodia and Laos	187
9.	Seveso	197
10.	Love Canal	229
11.	Where Do We Go from Here?	237
Index		245

Introduction

On 10 July 1976 a chemical reactor near the Italian town of Seveso overheated, discharging its contents over the town and its environs. The chemical cocktail vented from the reactor contained, among other things, an extremely toxic chemical 2,3,7,8-tetrachlorodibenzo-p-dioxin, more simply known as dioxin. When eventually it became known that the substance had contaminated some hundreds of acres of residential property and farmland, there were fears for the health of local residents.

This concern was justified. Even very small doses of dioxin have been shown to be lethal to animals; and concentrations of the order of several millionths of a gram will cause them to give birth to malformed offspring. Furthermore the chemical has now been shown to cause cancer in animals.

Because of these known effects there was much speculation about the immediate and longer-term health risk for the residents in the weeks following the Seveso accident. Four years later there is still speculation, in spite of the evidence supplied from the many other incidents involving exposure to dioxin which have come to light. The majority of these have occurred in industry, during the manufacture of 2,4,5-trichlorophenol—the starting point for the synthesis of the herbicide 2,4,5-T (2,4,5-trichlorophenoxyacetic acid) and for the antibacterial agent hexachlorophene.

The first recorded exposure of workers to dioxin in industry occurred over 30 years ago. Since then there have been over 20 further such industrial accidents. In spite of this situation, the long-term risks of exposure to the chemical are still not clear. It required the accident at Seveso and its attendent publicity to galvanize industry and government regulatory authorities into action. Investigations were begun into the health of dioxin-exposed workers in industry.

Seveso was not the first occasion on which a "civilian" population was affected by dioxin. Many people in Vietnam are also known to have been exposed to the chemical through the widespread military application of the herbicide "Agent Orange" by U.S. forces during the Vietnam War. Agent Orange was heavily contaminated with dioxin. So dioxin contamination has occurred in widely differing situations—in industry and in the community at large; in peaceful and in military contexts. This book describes incidents in all these categories. I have attempted to discuss each event in sufficient depth to give as comprehensive a picture as possible of the repercussions, human and environmental, of the release of dioxin. Inevitably some chapters are more detailed than others—a suitable reflection of the information available. Two chapters in particular—those describing the situation at Seveso and the effects of herbicide spraying in Vietnam—are the most detailed. This was a deliberate decision. In both cases large populations were affected by exposure to a toxic chemical and there are obvious lessons to be drawn from both events.

Vietnam and Seveso represent two aspects of a similar problem, that is, the consequences of long-term exposure to low levels of dioxin. Assessing the consequences of this exposure for the health of the people of Vietnam has proved to be a daunting task. In Vietnam, dioxin exposure occurred in the middle of a devastating war. Millions of people were uprooted as a result of the fighting and it is unlikely that everyone who was exposed to Agent Orange will be traced. For those who have been located, is there evidence that the herbicide damaged their health? Vietnamese scentists and U.S. veterans of the Vietnam War say yes, unequivocally. Other scientists still have doubts, however. Why should there be such disagreement? The chapter on Vietnam and 2,4,5-T attempts to answer this question by unraveling the story behind the development and use of 2,4,5-T, and by analyzing the case which has been made both for and against the use of the herbicide.

At Seveso, residents of the town were exposed to dioxin for two weeks before Italian officials realized the seriousness of the situation and ordered an evacuation. Why the delay? The officials blame the company responsible for the accident in which dioxin was released. The company, they say, did not explain the dangers involved until two weeks after the accident. In defense, the company argues that it took two weeks to assess the extent of the problem; and when this became clear, it actually demanded that the residents be evacuated.

Once the problem at Seveso was defined, the population was subjected to a battery of medical investigations to assess the consequences of its exposure to dioxin. The chapter on Seveso discusses these programs, the findings, and the limitations of the various studies. Communication between residents, scientists, clinicians, and the local authorities at Seveso was a problem from the outset and left many people disillusioned about the whole medical surveillance program.

It has been argued that an accident like Seveso, and particularly its aftermath, would never occur in other industrial countries. But this, of course, is just speculation. Accidents involving dioxin have occurred in

Introduction 3

many countries, and while each accident may be unique, there are problems common to all. The accidents involving the discharge of dioxin are by now reasonably well known. But what about other chemical processes? Accidents happen with these too. The consequences of these accidents are much more serious if chemical factories are sited in the middle of a residential area like Seveso. Planning to prevent a disastrous accident or to mitigate the effects of one is a constant headache for local authority planners. Seveso has implications for most town planners and provides some useful lessons.

Issues of secrecy and withholding of information are also part of the dioxin story. An accident at the British company Coalite and Chemical Products Limited in 1968 is discussed at some length in this book to illustrate the problems which arise when companies are not forthcoming about subjects which are of public interest. In recent years—and particularly since the accident at Seveso—Coalite has become far more secretive about medical surveillance of its dioxin-exposed workforce. In consequence the company has had a bad press. But it is not only its relationship with the media which has suffered; Coalite is also at odds with local planning authorities over the issue of disposal of dioxin-contaminated equipment and waste. There is a public relations message in the Coalite chapter for many industrial companies.

The chapters on Cambodia and on Love Canal are concerned with a mystery and a problem for the future, respectively. In 1969 Cambodia was desperately striving to preserve its neutrality at the same time that its neighbors Vietnam and Laos were embroiled in a war. Suddenly in the Spring of 1969 an area in the southwest of the country was sprayed with herbicides. The spraying was deliberate. But who did it remains a mystery, as does the reason for the attack. Equally intriguing, however, is the allegation, widely publicized, that this area was deliberately bombed by the U.S. in the Spring of 1969. There is ample evidence to prove that herbicide spraying took place in Cambodia in 1969, but there is no on-the-ground evidence to support the bombing thesis.

As for Love Canal this exemplifies the danger posed by toxic waste. This is very much a problem for the future. Toxic waste is likely to increase. Its safe disposal is becoming an issue which requires urgent resolution. Love Canal is an example of what not to do.

In preparing this book, I envisaged its being read by the lay reader, as well as the scientist with an interest in the subject. In consideration of both viewpoints, I have written most chapters in a manner which I trust is intelligible to most readers. For the reader with more specialized knowledge there are five chapters discussing the chemistry of dioxins, their biological or toxic properties, their effect on the skin, as well as alternatives to the herbicide and bacteriacide which contains dioxin impurities.

There is more than one dioxin. In fact there are 75 different types, all of which are referred to in the chapter discussing the chemistry of the subject. Each has its own specific chemical name. For obvious reasons, the abbreviated form of "dioxin" is used throughout the rest of the book. In this case dioxin refers to the most toxic of the 75 different chemicals in this class of compounds, 2,3,7,8-tetrachlorodibenzo-*p*-dioxin.

A knowledge of the chemistry of dioxins leads to a greater understanding of the issues involved in the field. For that reason the chapter on chemistry is the first in the book. It is followed by the other technical chapters on toxicology, skin problems, and alternatives. The reader who is not interested in the detailed properties of the chemicals, but more concerned to know about their use and their effect on humans, may start at Chapter 6.

1
Chemistry and Occurrence of Dioxins

The abbreviation "dioxin" has become a common term for anyone referring to the polychlorinated dibenzo-*p*-dioxins (PCDDs). More often than not "dioxin" has been used as an abbreviation for the dibenzodioxin with four chlorine atoms 2,3,7,8-tetrachlorodibenzo-*p*-dioxin (TCDD). The terminology for TCDD is quite simple. "Tetrachloro" refers to the four chlorine atoms on the two benzene rings. "Dibenzo" refers to these two benzene molecules, and dioxin describes the two oxygen bridges holding the two benzene rings together (as shown in Figure 1). The designation *p*, or para, refers to the position of the oxygen atoms in relation to the benzene molecules. The number of PCDD isomers varies depending on how many chlorine atoms are attached to the parent molecule. Four chlorine atoms attached at random on any of the eight positions labeled in Figure 1 confers the greatest degree of flexibility, and hence most isomers occur in the tetrachlorinated group (see Table I).[1]

Both the position and number of chlorine atoms attached to the dibenzo-*p*-dioxin ring structure are important in determining the toxicity of the PCDDs. The symmetrically substituted 2,3,7,8-tetrachlorinated dibenzo-*p*-dioxin (2,3,7,8-TCDD) is by far and away the most toxic of the PCDDs.[2] When toxicity is assessed by the standard LD_{50} test—which measures the quantity of substance required to kill half the experimental animals within a certain group—2,3,7,8-TCDD is 1000–10,000 times more toxic than an isomer with three chlorine atoms, 2,3,7-trichlorodibenzo-*p*-dioxin or one with two chlorine atoms, 2,8-dichlorodibenzo-*p*-dioxin.[2] The lateral positions 2,3,7,8 of the dioxin ring structure are the most important as far as toxicity is concerned, and all four positions need to be filled for maximum toxicity. Adding another chlorine atom to 2,3,7,8-TCDD to produce 1,2,3,7,8-pentachlorodibenzo-*p*-dioxin leads to a reduction in toxicity. The new compound is only half as toxic as 2,3,7,8-TCDD.[2]

Not all of the polychlorinated dibenzo-*p*-dioxins have been synthesized. A Swedish chemist, Professor Christopher Rappe of the University of Umea, has prepared and identified 31 of the possible 75 isomers,[3] far more than any other investigator in the field has prepared so far.

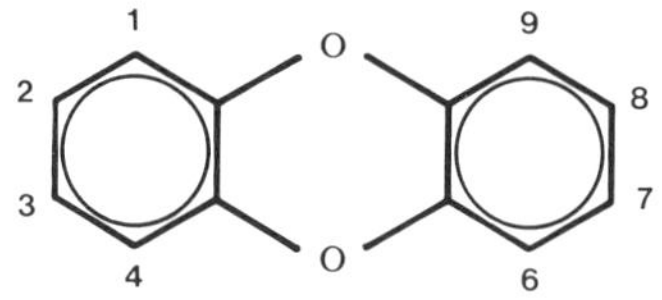

FIG. 1. Dibenzo-p-dioxin.

Polychlorinated dibenzo-p-dioxins are the by-products of other manufacturing processes. The principal source of the PCDDs is thought to be in the production of chlorinated phenols, the annual worldwide production of which is estimated to be about 150,000 tons. Chlorinated phenols are used directly or indirectly as slimicides, herbicides, fungicides, and bacteriacides. Methods for synthesizing chlorinated phenols vary depending on the product required. PCDDs have not been detected in every chlorinated phenol, and this may either be due to the method of synthesis of the phenol or to the number of chlorine atoms present. The principal method of making chlorinated phenols is by the direct chlorination of phenol; 2,4-dichloro, 2,4,6-trichloro, 2,3,4,6-tetrachloro, and pentachlorophenol are all prepared in this way (see Table II for their structures). Contaminating PCDDs have been observed in the last three.[4]

The second method of synthesis of chlorinated phenols is exclusive to the production of 2,4,5-trichlorophenol. The process involves the alkaline hydrolysis of 1,2,4,5-tetrachlorobenzene in the presence of either meth-

TABLE I. Number of Isomers Possible with Varying Degree of Chlorination (Rappe, 1978)[a]

Number of chlorine atoms	Number of polychlorinated dibenzo-p-dioxin isomers
1	2
2	10
3	14
4	22
5	14
6	10
7	2
8	1
Total	75

[a]Reference 1.

TABLE II. Types of Chlorinated Phenol

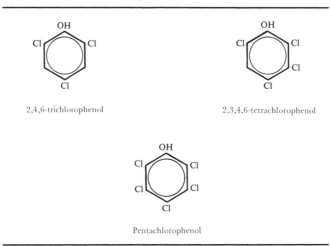

anolic sodium hydroxide at elevated pressures or ethylene glycol and sodium hydroxide at atmospheric pressure (see Figure 2).

In the course of producing 2,4,5-trichlorophenol, the toxic contaminant 2,3,7,8-tetrachlorodibenzo-*p*-dioxin is formed. The formation of the dioxin is unavoidable and it is brought about by the condensation of two molecules of 2,4,5-sodium trichlorophenate (see Figure 3). Normally, contamination of the trichlorophenol mix with 2,3,7,8-TCDD is minimal. However, should the solvent be distilled off and the temperature allowed to rise, an exothermic reaction is initiated. The exothermic reaction which starts at a temperature of about 230°C rises rapidly to 410°C and is attributed to the decomposition of sodium-2-hydroxyethoxide (NaOCH$_2$—CH$_2$OH).[5] This ethoxide is the sodium salt of the ethylene glycol solvent used in the mix. Increasing temperature is reported to favor production of PCDDs and, indeed, it has been shown that the TCDD concentration of the trichlorophenol mix increases with increasing temperature.[6] According to

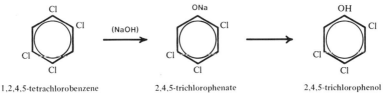

FIG. 2. Hydrolysis of tetrachlorobenzene to 2,4,5-trichlorophenol.

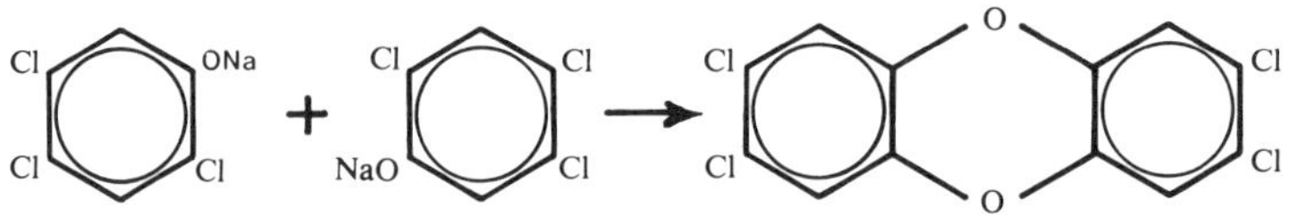

FIG. 3. Condensation of 2 molecules of sodium trichlorophenate to form 2,3,7,8-tetrachlorodibenzo-*p*-dioxin.

Dr. Jorg Sambeth of the Swiss-based chemical company Givaudan, it is now possible to predict the amount of dioxin produced in the trichlorophenol mix "fairly accurately" by simply knowing the final temperature.[7]

If PCDD formation is unavoidable in the manufacture of 2,4,5-trichlorophenol, so is the production of other toxic contaminants, in this case polychlorinated dibenzofurans (PCDFs). According to Rappe, in the production of 2,4,5-trichlorophenol both PCDFs and PCDDs are produced, the former at the start of the reaction and the latter at the end. Thus, says Rappe, at the start of the run to make this chlorinated phenol conditions would favor the production of 2,3,7,8-tetrachlorodibenzofuran (2,3,7,8-TCDF). The condensation reaction which produces the furan is referred to by chemists as a bimolecular process, i.e., two molecules are involved. The rate and extent of the reaction depend on the concentrations of the reactants involved. In the initial phase of the hydrolysis of tetrachlorobenzene only a few molecules of sodium trichlorophenate are present (see Figure 2). Should the temperature rise in the initial reaction stages, a condensation reaction involving tetrachlorobenzene and 2,4,5-sodium trichlorophenate would occur and 2,3,7,8-TCDF would be formed (see Figure 4). Two scientific groups working independently have reported that samples of Agent Orange—the herbicide used in Vietnam—and 2,4,5-T products currently available for sale all have PCDFs present in similar concentrations to the PCDDs.[8,9] According to one of the groups, both PCDFs and PCDDs are formed when 2,4,5-T is heated.[9]

Speaking at a meeting of the International Agency for Research on

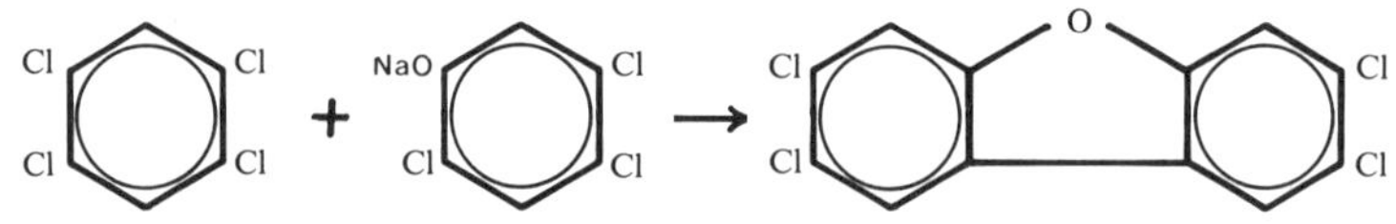

Tetrachlorobenzene   2,4,5-sodium trichlorophenate   2,3,7,8-tetrachlorodibenzofuran

FIG. 4. Condensation reaction involving tetrachlorobenzene and sodium trichlorophenate to form 2,3,7,8-tetrachlorodibenzofuran.

Cancer (January 10, 1978) Rappe[1] pointed out that where industrial accidents involving trichlorophenol plants had occurred, the nature of the toxic contaminants produced would depend on the stage at which the reaction had been interrupted. An accident at an early stage would favor PCDF formation, and at the end PCDD production. At Seveso, the accident in the trichlorophenol reactor occurred 6½ hours after the completion of the run. Little tetrachlorobenzene would be present at this stage; the main component of the reactor mix would be 2,4,5-sodium trichlorophenate. When the temperature rose, as it did at Seveso, the bimolecular reaction would involve the condensation of two molecules of sodium trichlorophenate to form 2,3,7,8-tetrachlorodibenzo-p-dioxin (Figure 3). Analysis of the soil at Seveso confirmed these theoretical predictions; 2,3,7,8-TCDD was the major contaminant present.

Although most concern has been expressed about the PCDDs present in trichlorophenol mixes, PCDFs have not passed unnoticed. The reason for this is easy to find; the chlorinated dibenzofurans may not be as toxic as their corresponding chlorinated dibenzodioxin analogs but they are, nevertheless, extremely toxic in their own right. It appears that like the chlorinated dioxins, the toxicity of the furans varies between animal species. According to Professor John Moore's group at the National Institute of Environmental Health Sciences in North Carolina, the toxicity of 2,3,7,8-TCDF in mice when measured using the LD_{50} test is two- to fourfold lower than the corresponding dioxin isomer 2,3,7,8-TCDD.[10]

Public anxiety about dioxins has forced many government regulatory authorities, and the U.S. Environmental Protection Agency in particular, to focus their attention on the herbicide 2,4,5-T. Indeed the EPA's ten-year campaign—begun in 1970—to ban this herbicide is testimony to its concern about the product.

Contamination of 2,4,5,-T with dioxin is due to poor industrial cleaning processes. The contamination occurs at an early stage in the preparation of the herbicide when 2,4,5-trichlorophenol is produced, and it is dioxin-tainted trichlorophenol which is subsequently converted to dioxin-contaminated 2,4,5-T. The conversion of 2,4,5-trichlorophenol to 2,4,5-T is usually a two-stage process. The chlorinated phenol is first reacted with chloroacetic acid in an alkaline medium to form the sodium salt of 2,4,5-T. This is acidified to form the herbicide itself (see Figure 5).

To produce longer-acting herbicides 2,4,5-T can be readily converted to an ester or amine salt by reacting with alcohols or amines, respectively. Esters can also be produced directly from the sodium salt of 2,4,5-trichlorophenol, as shown in Figure 6 where R represents an alkyl group.

In most countries, manufacturers of 2,4,5-T are required by law to keep dioxin levels below 0.1 ppm (parts per million). The limit in Britain is now ten times lower than this, i.e., 0.01 ppm. All manufacturers claim that

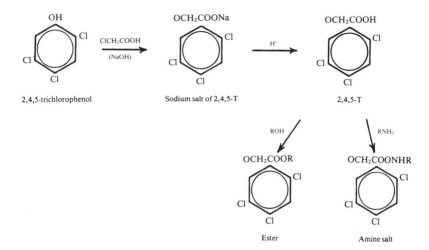

FIG. 5. Formation of 2,4,5-T and its ester and amine salt.

they adhere to the legal limit, and in most cases they police themselves. Few government agencies have the facilities to check these claims and they have often to be taken on trust.

In the United Kingdom, however, the Office of the Government Chemist (OGC) performs periodic checks on the dioxin level of commercial preparations of 2,4,5-T.[11] But in Britain there is not the same degree of concern about 2,4,5-T that there is in the United States. The reason for this may have something to do with the quantity of 2,4,5-T used in Britain, which is small, and much lower than the amounts used in the United States.[12]

Measurements of dioxin require the use of a gas chromatograph and mass spectrometer—the former to isolate the chemicals of interest, and the latter to identify them, much like taking a fingerprint. The spectrometer in use by scientists in the OGC, at least until 1980, was a Model MS 9. This was first manufactured in 1959 and it is many times less sensitive than the newer model, MS 50, now in use in the United States. The older MS 9

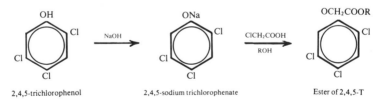

FIG. 6. Formation of 2,4,5-T ester.

could, in a pinch, measure dioxin levels at the parts-per-million level, whereas the later MS 50 will measure dioxin levels at the parts per trillion (U.S. trillion, i.e., a million million). The older MS 9 is a spectrometer a million times less sensitive than the model recognized to be standard equipment for any laboratory now concerned with dioxin measurements.

Givaudan, a subsidiary of Hoffman La Roche, uses trichlorophenol for the manufacture of hexachlorophene, a widely used bacteriacide. Hexachlorophene is prepared by the condensation of two molecules of 2,4,5-trichlorophenol with formaldehyde in the presence of concentrated sulfuric acid as shown in Figure 7.

The bacteriacide is used in the cosmetics industry as a preservative. For medical purposes hexachlorophene is widely used in the control of bacteria and particularly staphylococcal organisms. In view of its use in hospitals, every effort is made to ensure that hexachlorophene is free of contaminating substances. In theory any dioxins present ought to be removed in the additional purification steps 2,4,5-trichlorophenol undergoes before conversion to hexachlorophene. The reality is slightly different and the bacteriacide is still contaminated. Rappe reports that TCDD has been measured in hexachlorophene at a level of 0.03 ppm.[4]

Brief reference has been made to chlorinated phenols other than 2,4,5-trichlorophenol. According to Rappe,[4] commercial sources of pentachlorophenol on sale in Switzerland were found to contain both PCDDs and PCDFs with the former present in the greatest quantity. The dioxins in pentachlorophenol have been identified and those with six, seven, and eight chlorine atoms shown to predominate.[13,14] The toxic effects of technical and pure grade pentachlorophenol on the liver of rats cannot be explained by the phenol itself, and it has been suggested that the effects are more consistent with its contamination by PCDDs and PCDFs.[15]

Dioxins and furans have been observed too in commercial preparations of 2,4,6-trichlorophenol and 2,3,4,6-tetrachlorophenol on sale in Scandinavia. The concentration of the dioxins present in these phenols is only about 1/20th of the furan concentration. Although it is known that there are some ten different furan isomers in these products, there is little

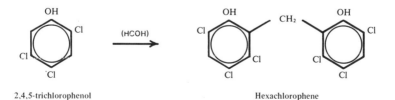

FIG. 7. Formation of hexachlorophene.

information available on the identity and quantity of the individual isomers.[14,15]

Another source of the furans is the polychlorinated biphenyls (PCBs). In 1970, Dutch scientists reported that PCBs available in Europe were contaminated with the furans at the ppm level.[16] But more important, the group reported that the toxic effects observed to occur with the PCBs were almost certainly due to their contamination with furans.[16] This observation was made by Dr. J. G. Vos and his colleagues of the National Institute of Public Health at Bilthoven in the Netherlands. In many ways it was a chance discovery. Vos was interested in the toxic effects of a number of PCBs given the trade names Clophen A 60, Phenoclor DP 60, and Aroclor 1260. His findings were quite unexpected. Anticipating all three to have similar toxic properties, Vos found that this was not the case. Both Clophen and Phenoclor were far more toxic than Aroclor. Clophen and Phenoclor caused death, liver necrosis, and subcutaneous and abdominal water retention in chickens; such findings were, however, rarely seen with Aroclor.[16,17] Vos and his colleagues concluded that those differences were attributable to the presence of other more toxic compounds. Further analysis confirmed this, and the presence of tetra- and pentachlorodibenzofurans in Clophen and Phenoclor was firmly established.

Contamination of PCBs with furans is not, however, a subject of purely academic interest. For 1684 people in Japan it is a major health problem. The 1684 were victims of a poisoning episode involving cooking oil contaminated with PCBs and polychlorinated dibenzofurans.

The poisoning was inadvertent and, like other episodes documented in this book, identifying its cause was not straightforward. That there was a serious problem became obvious following an epidemic of chloracne in March 1968 in people residing in the Fukoaka prefecture (see Chapter 5 for details of the skin disease chloracne). March was the first month in which chloracne was observed. The problem worsened in successive months as increasing numbers of chloracne cases attended outpatient dermatology clinics. By June the number of new cases had reached its peak. For the next two months the problem remained acute but by September the outbreak seemed to have been contained and the number of new cases began to fall. The outpatient dermatology clinic at Kyushu University recorded its last two new cases in January 1979; this one clinic alone had seen 138 chloracne cases since March of the previous year.

To investigate the problem a study group was formed at Kyushu University in October 1968. Included in the group were members of the Departments of Medicine, Pharmaceutical Sciences, and Technology. At the outset of the study two facts were clear and suggested that the cause of the problem might in fact be easier to trace than the clinicians had initially indicated. By October it was obvious that the outbreak of chloracne cases was restricted to a certain period of 1968 and that the outbreak had no relation to any industrial occupation. The victims in all cases were whole families and the breadwinners of each family were employed in a variety of occupations. From these observations, together with the knowledge that chloracne—as described for industrial cases—was restricted to compounds which were soluble in oil, the study group had to conclude that the causative agent was an oil-based product common to every individual's diet. Rice-bran oil was quickly isolated as the common factor in all cases; samples of oil were analyzed and a PCB (Kanechlor 400) found to be present as a contaminant. With oil the cause of the problem, the group became known as the Yusho (literally meaning oil disease) cases.

The canned rice oil thought to be the cause was traced to a shipment of oil despatched on 5 February 1968. Analysis of the oil had shown toxic metals such as copper, nickel, zinc, cobalt, arsenic, mercury, or pentachlorophenol to be absent, but polychlorinated biphenyls to be present in high concentration of the order of 2000-3000 ppm.[19] The PCBs found in the oil were characteristic of a particular type, Kanechlor 400, produced by the Kanegafuchi Chemical Industry Company. But Kanechlor 400 was not only measured in the oil, it could be detected in tissue samples from the patients. A variety of tissues were examined from live and recently deceased patients. The tissues included heart, marrow, small intestine, adipose fat, subcutaneous fat, trachea, and even fetal tissue; evidence of the PCBs was present at all times.[20-22]

To confirm the PCB poisoning, some 109 samples of bottled rice oil shipped between October 1967 and October 1968 were analyzed. The PCBs were found to be confined to oil shipped in the 12-day period 7-19 February 1968. No samples of oil were available from the company for the two preceding days, the 5th and 6th of February 1968. This intensive investigation of the chemistry of the oils was fairly conclusive, and the findings matched perfectly with the results of the epidemiological studies; the chloracne cases were confined to users of oil despatched in the two-week period in February 1968. Subsequent analysis of the oil confirmed the presence of high concentrations (5 ppm) of polychlorinated dibenzofurans. The furans seen were the tetra (four chlorine atoms) and pentachlorofurans (five chlorine atoms) and considered to be the most toxic of this class of compounds.[19] It goes without saying that many observers, including the doctors monitoring the condition of their patients with chloracne, suspect that the toxic furans are the real cause of the public health problem for the 1684 people concerned. (Chapter 4 has further details about the health status of the Yusho patients.)

There are still the questions why and how did the furan-laced PCB contaminate the cooking oil? Kanechlor 400 was used to heat the cooking oil under reduced pressure to remove any unwanted odors from the final product. The PCB heating oil must, therefore, have leaked into the cooking oil at some stage of the processing, a theory which gained credence when small holes were later discovered in an old section of the internal heating pipe.[19] The occurrence of the furans in PCBs is attributed to heating of the polychlorinated biphenyls. It has been shown recently that pyrolysis of technical grade PCB mixtures will yield some 30 major and 30 minor polychlorinated dibenzofurans.[4] The most toxic of the furans, the tetrachlorinated isomer 2,3,7,8-TCDF (the furan analog of 2,3,7,8-tetrachlorodibenzo-*p*-dioxin) is one of the major components seen after heating the PCB mixture. Yield of the furans after heating the PCBs is also high,[23,24] and Professor Rappe considers, therefore, that uncontrolled burning of the polychlorinated biphenyls can be an important environmental source of these compounds.[4] Rappe suggests that the reaction scheme in Figure 8 explains what happens when a PCB is heated. As proof of this theory Rappe and his colleagues have observed two of the most toxic furan isomers in fat obtained from a Gray Seal. The authors report that 2,3,7,8-TCDF and 2,3,4,7,8-pentachlorodibenzofuran were present in the fat and were probably due to direct contamination by PCBs.[25]

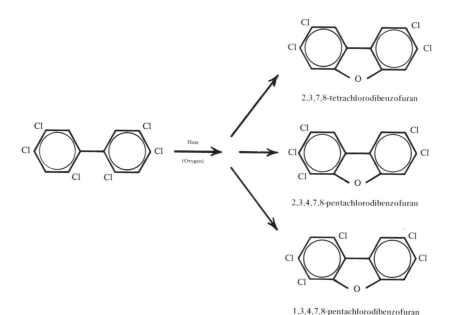

FIG. 8. Pyrolysis of hexachlorobiphenyl.

Another major source of polychlorinated dibenzodioxins and polychlorinated dibenzofurans is the flyash and flue gases of municipal incinerators.[26,27] Concentrations of dioxins and furans at 0.2 and 0.1 ppm, respectively, have been measured in the flyash from a Swiss municipal incinerator; concentrations three times as high were recorded for the ash of a Swiss industrial heating facility.[23] A large number of different isomers—as many as 30–40 of dioxins and furans—have been detected in the ash. The highly toxic 2,3,7,8-tetra- and 1,2,3,7,8-pentachlorodibenzo-p-dioxin are only minor constituents of the ash, whereas the highly toxic furan isomers are the major constituents.[23,28]

A report produced by the Dow Chemical Company, however, comes to quite different conclusions about the quantity of toxic dioxins in flyash. Dow, as the major manufacturer of the herbicide 2,4,5-T in the United States, claims that the herbicide is not the major source of dioxins detected in the environment. In the report "The Trace Chemistries of Fire"[29] prepared by Dow for the State of Michigan's Department of Natural Resources (DNR), the company insists that dioxins are present in ash collected from chemical tar burners, fossil-fueled power plants, the mufflers of automobiles and trucks, household chimneys, and even charcoal-broiled steaks. Dioxins, according to Dow are "ubiquitous." Their presence, the company argues, is due to the existence of a natural phenomenon: "trace chemistries of fire." A natural phenomenon which will produce dioxins has obvious advantages for the company. If it were true, the finger could no longer be pointed at Dow as one of the major sources of dioxins contaminating the environment. Dow's latest report was produced for the DNR after Dow found that fish taken from the nearby Tittabawassee River contained measurable amounts of chlorinated dioxins and polychlorinated biphenyls. Effluent from Dow's huge Michigan complex is discharged directly into the river and, not unnaturally, the company was assumed to be the source of the dioxins. Dow disputed this charge; the company's report claimed that dioxins are produced in many combustion processes and are widespread. But the DNR was not convinced that dioxin production was so universal and the Department insisted on additional measurements. Dr. Robert Bumb, Director of Research at Dow's Michigan complex opposed this on the grounds that additional tests would not only be expensive, but of little value; he also claimed that other scientists were confirming Dow's discovery.[30]

This, however, is not quite true. Dow's "discovery" that dioxins are produced in commercial incinerators is not new; the two independent European groups referred to earlier both reported before Dow that the flyash of municipal and industrial incinerators contained polychlorinated dibenzo dioxins and polychlorinated dibenzofurans.[26,27] It was Dow's subsequent discovery that chlorinated dioxins were present in the ash collected from other combustion processes, which led them to develop their theory on the "trace chemistries of fire." There had to be a common factor to explain the occurrence of dioxin in so diverse a range of products as the ash from chemical tar burners and charcoal-broiled steaks. To explain the wide occurrence, Dow developed its theory on trace chemistries—defined as "numerous chemical reactions occurring during combustion at very low concentrations, parts per million and lower." Yields from these reactions are very low, of the order of $10^{-9}\%$.

The company attributes the formation of chlorinated dioxins to the presence of dioxin building blocks, which would include chlorine and chlorinated aliphatic and aromatic hydrocarbons. Metals which may be present—vanadium and nickel, for example—act as catalysts in a sea of chemical reactivity including pyrolysis, oxidation, reduction, and acidolysis. In a similar poetic vein, the report adds that in this sea, "ions, electrons, free radicals, free atoms and molecules form, combine and decompose. The process is repeated time and time again," and chlorinated dioxins, Dow suggests, must be formed in this process.

Not necessarily, says Dr. Christopher Rappe. Results obtained by him, a colleague, Dr. Hans Rudolf Buser, and a further collaborator, Dr. Hans Paul Bosshardt of the Swiss Federal Research Station, suggest that the dioxin precursors are not so nebulous. Rappe considers that chlorinated phenols and chlorinated diphenyl ethers (see Figure 9) are the main dioxin precursors in municipal incinerators.[31] When these two groups of compounds are heated under laboratory conditions, Rappe has found that they produce isomers of chlorinated dioxins in similar proportions to those seen in commercial incineration. Furthermore, Rappe points out that "chlorinated phenols and chlorinated diphenyl ethers yield PCDDs (polychlorinated dibenzodioxins) at a level exceeding the level found by Dow by a factor of 1,000,000. Consequently, says Rappe, there is good "evidence that these compounds are the main precursors" of dioxins.[31] Because Dow was a large-scale producer of chlorinated phenols, Rappe assumed that it would be interested in his results, and duly forwarded them to the company. It appears, however, that the company was either not interested in the findings, or that it chose to ignore them, for there are no analyses of these precursors in the Dow report.

There was also skepticism about the validity of the results reported by

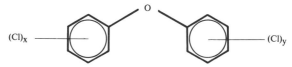

FIG. 9. Halogenated diphenyl ethers. Halogen can be chlorine or bromine. X and Y represent the number of halogen atoms which can be from 1–5 per benzene ring.

Dow in its report. According to Rappe, the chromatographic separation techniques used by Dow to identify dioxin isomers were poor and should have been reanalyzed using another method.[31] In its report, Dow published a chromatographic trace showing three small peaks for four dioxin isomers, each with four chlorine atoms. It bore little resemblance to a published method of Rappe and his colleagues in which ten of these tetrachlorinated dioxin isomers could be clearly identified (see Figure 10). It was Rappe's opinion that Dow had overestimated the quantity of the tetrachlorinated dioxin isomer (2,3,7,8-TCDD) which would be present in the flyash of municipal incinerators.[32]

2,3,7,8-TCDD is many times more toxic than any of the other chlorinated dioxin isomers studied so far, and Rappe said that, in European incinerators, it only represents some 1%–3% of the tetrachlorinated dioxin isomers in flyash. Dow, on the other hand, reported that in many of their

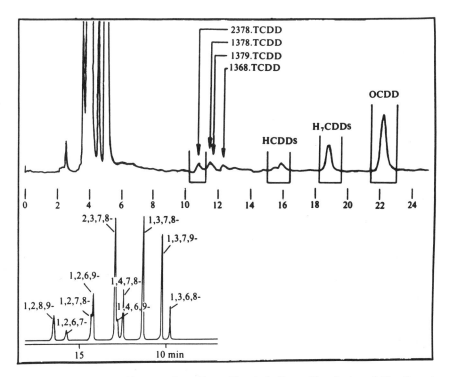

FIG. 10. Chromatographic trace from Dow Chemical, Trace Chemistries of Fire Report (top), and from Rappe and Buser method (bottom). [From Hay, A., Dispute over Dow Chemical's theory of dioxin traces, *Nature (London)* **281**, 619-620 (1979), courtesy of McMillan Journals Ltd.]

investigations 2,3,7,8-TCDD was the major isomer. Rappe suggested that this dramatic difference between his and Dow's results could be explained by the fact that Dow analyzed samples from one of its own incinerators. Dow incinerates residue from the reactor used to make 2,4,5-trichlorophenol, the precursor of the herbicide, 2,4,5-T. As the incineration of 2,4,5-trichlorophenol, and of the sodium salt in particular, is known to produce 2,3,7,8-TCDD, Rappe suggested that the chlorinated phenol may have already been present in the reactor.

Rappe and his colleagues have so far prepared and identified 31 of the 75 chlorinated dioxin isomers which are considered to be theoretically possible. Although Dow used a far fewer (and unstated) number in its study, it believed that it had identified enough isomers to make the report valid. In addition, Bumb pointed out that[33] another Dow scientist, Dr. D. E. Townsend, had verified the trace chemistries theory from thermodynamic principles. Bumb said that Townsend, in an unpublished report, found a striking correlation between observed and predicted values for the dioxin isomers.[34] Rappe, on the other hand, disputed this saying that Townsend's principal assumption that there is a constant ratio between dioxin isomers with different numbers of chlorine atoms was not borne out in practice, and "consequently his theory is wrong."[31] More recently Rappe told the author that analyses of tissue samples from animals and samples from other environmental sources had shown that there is a general background level of PCDDs, and more often at a somewhat higher background level of PCDFs. But, says Rappe, there are also environmental blanks. For this reason Rappe remains skeptical about Dow's hypothesis that every incineration process will yield dioxins.[35]

Dow appeared to be unaffected by the skepticism with which its results had been received by scientists in Europe and the United States. Bumb claimed that "results which depart from traditional and commonly held beliefs routinely provoke skepticism."[33] And he remained confident about Dow's results and the company's conclusions.

Bumb's optimism was justified. Recent research has indeed shown that there are other precursors of dioxins in incinerators. According to Dr. Otto Hutzinger of the University of Amsterdam, polyvinylchloride (PVC) and lignin are two compounds which on burning could form dioxins. It is conceivable that inorganic chlorine might be available in incinerators for the chlorination of dibenzodioxins. Dow scientists were the first to claim that chlorine from both inorganic and organic sources could be used in the formation of polychlorinated dibenzodioxins.

The Dow scientists published their trace chemistries of fire theory as a formal scientific paper in the American journal *Science* in October 1980. It is of interest to note that the results published by Dow are now similar to those obtained in European laboratories. Although Dow claims that the

2,3,7,8-TCDD isomer is present in significant quantities in the flyash, the results indicate that it accounts for less than 1% of the total PCDD isomers present. According to Dow, octachlorodibenzo-*p*-dioxin is present in the ash in the greatest concentrations, followed closely by heptachlorinated PCDD isomers. Hexachlorinated dioxins represent about 8% of the total PCDD compounds, the Dow scientists claim. Italian and Swiss scientists have reported similar findings.[37,38]

Contamination of flyash by PCDDs is clearly a problem and the question arises "Can anything be done to reduce the amount of dioxins produced in large incinerators?" Apparently it can. There are two alternatives: either to reduce the amount of organic chlorinated compounds burned in incinerators, or to increase the efficiency of incineration. Scientists at the University of Amsterdam have noted that PCDD content of flyash from incinerators in Holland was 20 times higher than the concentration in flyash from the Czechoslovakian town of Bratislava where far less PVC is incinerated.[39] As for improvements in combustion efficiency, Dr. Warrent Crummett, a well-known chemist working for the Dow Chemical company, claims that by using extra fuel in a company incinerator octachlorodibenzodioxin concentration in flyash fell dramatically and no 2,3,7,8-TCDD was detectable.[40]

There are sources of dioxins other than those referred to above. For example, it is known that when vegetation sprayed with 2,4,5-T is burned 2,3,7,8-TCDD is formed. The amount of dioxin produced, however, is small.[41,42] According to scientists at Dow Chemical and the authors of one of the reports on dioxin formation from the burning of 2,4,5-T sprayed vegetation, the burden added to the environment by this process "is no larger than 1 part per trillion of (2,3,7,8)-TCDD per ppm of 2,4,5-T residue burned." In other words, the amount of dioxin formed is of the order of 1 part per million. And this level according to a review panel of the Australian National Health and Medical Council Research Council does not constitute a health hazard.[43]

Finally there are the effects of light on dioxins and their precursors. One potential source of dioxin formation is the closure of the ring system of chlorinated *o*-phenoxyphenols (see Figure 9) by a photochemical effect induced by light. These phenoxyphenols, the so-called predioxins,[44] are very common impurities in commercial chlorophenols, present in concentrations as high as 1%–5%.[45]

Dioxins could conceivably act as their own precursors by another photochemical process, dechlorination. Chlorine atoms can be selectively removed from polychlorinated dioxins, thus changing a dioxin molecule with six chlorine atoms to one with five, four, or even less chlorine atoms. In this way the less toxic hexa- and pentachlorinated dioxins might be converted to the toxic tetrachlorinated isomers. To test the theory the octa-

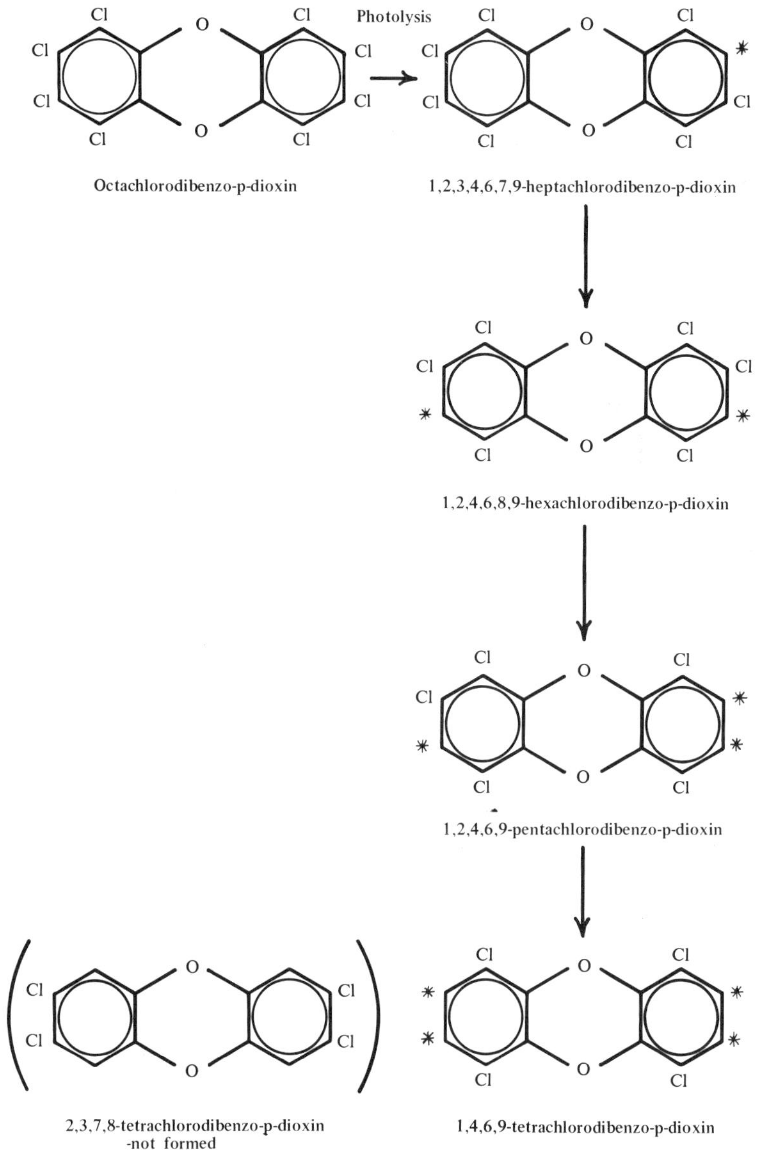

FIG. 11. Photochemical dechlorination of octachlorodibenzo-*p*-dioxin.

chlorinated dioxin was subjected to photolysis and the various isomers separated and identified. The major isomers identified are shown in Figure 11. It is quite clear from this process that the chlorine atoms removed from the polychlorinated dioxins by photolysis are the ones in the lateral positions (see Figure 11). The toxicity of the polychlorinated dioxins is due to the presence of chlorine atoms in the lateral positions. The toxic 2,3,7,8-tetrachlorodibenzo-*p*-dioxin is thus not likely to be formed by photochemical dechlorination of higher chlorinated dioxins.[46]

References

1. Rappe, C. Chemical aspects of polychlorinated dibenzodioxins (PCDDs) and polychlorinated dibenzofurans (PCDFs), presented at International Agency for Research on Cancer, January 10–11, 1978.
2. McConnell, E. E., Moore, J. A., Haseman, J. K., and Harris, M. W. The comparative toxicity of chlorinated dibenzo-*p*-dioxins in mice and guinea pigs, *Toxicology and Applied Pharmacology* **44**, 335–356 (1978).
3. Rappe, C. Personal letter, 30 July 1979.
4. Rappe, C., Buser, H. R., and Bosshardt, H. P. Dioxins, dibenzofurans and other polyhalogenated aromatics; production, use, formation and destruction, *Annals of the New York Academy of Sciences* **320**, 1–18 (1979).
5. Milnes, M. H. Formations of 2,3,7,8-tetrachlorodibenzodioxin by thermal decomposition of sodium trichlorophenate, *Nature (London)* **232**, 395–396 (1971).
6. Sambeth, J. Personal communication to author, reported in Hay. A. Seveso: the crucial questions of reactor safety, *Nature (London)* **281**, 521 (1979).
7. Sambeth, J. Personal communication to author reported in Hay, A. Seveso: No answers yet. *Disasters* **2**, 163–168 (1978).
8. Huckins, J. N., Stalling, D. L., and Smith, W. A. Foam-chemical chromatography for analyses of polychlorinated dibenzodioxins in Herbicide Orange, *Journal of the Association of Official Analytical Chemists* **61**, 32–38 (1977).
9. Ahling, B., Lindskog, A., Jansson, B., and Sundstrom, G. Formation of polychlorinated dibenzo-*p*-dioxins and dibenzofurans during combustion of a 2,4,5-T formulation, *Chemosphere* **6**, 461–466 (1977).
10. Moore, J. A., McConnell, E. E., Dalgard, D. W., and Harris, M. W. Comparative toxicity of three halogenated dibenzofurans in guinea pigs, mice and Rhesus monkeys, *Annals of the New York Academy of Sciences* **320**, 151–163 (1979).
11. Agriculture and Pollution: The Royal Commission on Environmental Pollution, 7th Report, September 1979 (Chairman, Sir Hans Kornberg).
12. Hay, A. Dioxin source is safe, *Nature (London)* **274**, 526 (1978).
13. Buser, H. R., and Bosshardt, H. P. Determination of polychlorinated dibenzo-*p*-dioxins and dibenzofurans in commercial pentachlorophenols by combined gas chromatography-mass spectrometry, *Journal of the Association Official Analytical Chemists* **59**, 562–569 (1976).
14. Rappe, D., Marklund, S., Buser, H. R., and Bosshardt, H. P. Formation of polychlorinated dibenzo-*p*-dioxins (PCDDs) by burning and heating chlorophenates, *Chemosphere* **7**, 269–281 (1978).

15. Goldstein, J. A., Friesen, M., Linder, R. E., Hickman, P., Hass, J. R., and Bergman, H. Effects of pentachlorophenol on hepatic drug metabolizing enzymes and porphyria related to contamination with chlorinated dibenzo-p-dioxins and dibenzofurans, *Biochemical Pharmacology* **26**, 1549–1557 (1977).
16. Vos, J. G., Koeman, J. H., van der Maas, H. L., ten Noever de Brauw, M. C., and de Vos, R. H. Identification and toxicological evaluation of chlorinated dibenzofuran and chlorinated naphthalene in two commercial polychlorinated biphenyls, *Food and Cosmetic Toxicology* **8**, 625–633 (1970).
17. Vos, J. G., and Koeman, J. H. Comparative toxicologic study with polychlorinated biphenyls in chickens with special reference to porphyria, edema formation, liver necrosis, and tissue residues, *Toxicology and Applied Pharmacology* **17**, 656–668 (1970).
18. Niguchi, K. Outline in *PCB Poisoning and Pollution*, ed. Kentaso Higuchi, Kodansha Ltd., Tokyo, Academic Press, New York (1976), pp. 3–7.
19. Duratsune, M. Epidemiologic Studies on Yusho, in *PCB Poisoning and Pollution*, ed. Kentaso Higushi, Kodansha Ltd., Tokyo, Academic Press, New York, 1976, pp. 9–23.
20. Tsukamoto, H. The chemical studies on detection of toxic compounds in the rice bran oils used by the patients of Yusho, *Fukuoka Acta Med.* **60**, 496–512 (1969).
21. Kojima, T. Chlorobiphenyls in the sputa and tissues, *Fukuoka Acta Med.* **62**, 25–29 (1971).
22. Kikuchi, M., Mikagi, Y., Hashimoto, M., and Kojima, T. Two autopsy cases of chronic chlorophenyls poisoning, *Fukuoka Acta Med.* **62**, 89–103 (1971).
23. Buser, H. R., Bosshardt, H. P., Rappe, C., and Lindahl, R. Identification of polychlorinated dibenzofuran isomers in flyash and PCB pyrolyses, *Chemosphere* **7**, 419–429 (1978).
24. Buser, H. R., Bosshardt, H. P., and Rappe, C. Formation of polychlorinated dibenzofurans (PCDFs) from the pyrolysis of PCBs, *Chemosphere* **7**, 109–119 (1978).
25. Rappe, C., Buser, H. R., Stalling, D. L., Smith, L. M., and Dougherty, R. C. Identification of polychlorinated dibenzofurans in environmental samples, *Nature (London)*, **292**, 524–526 (1981).
26. Olie, K., Vermeulen, P. L., and Hutzinger, O. Chlorodibenzo-p-dioxins and chlorodibenzofurans are trace components of flyash and flue gas of some municipal incinerators in the Netherlands, *Chemosphere* **6**, 455–459 (1978).
27. Buser, H. R., and Bosshardt, H. P. Polychlorieste dibenzo-p-dioxine, dibenzofurane und benzole in der asche dommunaler und industrieller verbrennungsanlagen, *Mitteilungen aus dem Gebiete der Lebensmitteluntersuchung und Hygiene* **69**, 191–199 (1978).
28. Buser, H. R., Bosshardt, H. P., and Rappe, C. Identification of polychlorinated dibenzo-p-dioxin isomers found in flyash, *Chemosphere* **7**, 165–172 (1978).
29. "The Trace Chemistries of Fire," a source of and routes for the entry of chlorinated dioxins into the environment by the Chlorinated Dioxin Task Force, the Michigan Division, Dow Chemical USA (November 1978).
30. Rawls, R. Dow finds support, doubt for dioxin ideas. *Chemical and Engineering News*, 12 February 1979, pp. 23–27.
31. Rappe, C. Comments on the Dow Report "The Trace Chemistries of Fire," Letter to Author, 26 March 1979.
32. Rappe, C. Reported in Hay, A., Dispute over Dow Chemical's theory of dioxin traces, *Nature (London)* **281**, 619–620 (1979).
33. Bumb, R. R. Personal letter, 7 June 1979.
34. Townsend, D. E. "The Environmental Background: the Fate and Behaviour of Trace Chemicals formed by Combustion," Dow Chemical Company 1978/1979.
35. Rappe, C. Personal letter, 16 June 1981.
36. Bumb, R. R., Crummett, W. B., Cutie, S. S., Gledhill, J. R., Hummel, R. H., Kagel, R. O., Lamparski, L. L., Luoma, E. V., Miller, D. L., Nestrick, T. J., Shadoff, L. A., Stehl, R. H.,

and Woods, J. S. Trace chemistries of fire: a source of chlorinated dioxins, *Science* **210**, 385–390 (1980).
37. Liberti, A. Formation of organochlorine compounds in incineration processes, presented at workshop on "Impact of Chlorinated Dioxins and Related Compounds on the Environment," Rome, 22–24 October 1980. Also reported in Hay, A., Chlorinated dioxins and the environment, *Nature (London)* **289**, 351–352 (1981).
38. Buser, H. R. High resolution gas chromatography of 22 Tetrachlorinated-p-dioxin (TCDD) isomers, presented at workshop on "Impact of Chlorinated Dioxins and Related Compounds on the Environment," Rome, 22–24 October 1980.
39. Olie, K., Lustenhouwer, J. W. A. and Hutzinger, O. Chlorinated dibenzo-p-dioxins and related compounds in incinerator effluents, presented at workshop on "Impact of Chlorinated Dioxins and Related Compounds on the Environment," Rome, 22–24 October 1980.
40. Crummett, W. B., Bumb, R. R., Lamparski, L. L., Mahle, N. H., Nestrick, T. J. and Whiting, L. Environmental chlorinated dioxins from combustion—the trace chemistries of fire hypothesis, presented at workshop on "Impact of Chlorinated Dioxins and Related Compounds on the Environment," Rome, 22–24 October 1980.
41. Stehl, R. H. and Lamparski, L. L. Combustion of several 2,4,5-trichlorophenoxy compounds: formation of 2,3,7,8-tetrachlorodibenzo-p-dioxins, *Science* **197**, 1008–1009 (1977).
42. Ahling, B., Lindskog, A., Jansson, B., and Sundstrom, G. Formation of polychlorinated dibenzo-p-dioxins and dibenzo furans during combustion of a 2,4,5-T formulation, *Chemosphere* **6**, 461–466 (1977).
43. Press statement, Australian National Health and Medical Research Council, 16 June 1978. See also Hay, A., Dioxin source is safe, *Nature (London)* **274**, 526 (1978).
44. Nilsson, C. A., Andersson, K., Rappe, C., and Westermark, S. O. Chromatographic evidence for the formation of chlorodioxins from chloro-2-phenoxyphenols, *Journal of Chromatography* **96**, 137–147 (1974).
45. Nilsson, C. A., Norstrom, A., Andersson, K., and Rappe, C. Impurities in commercial products related to pentachlorophenol, in *Pentachlorophenol Chemistry, Pharmacology and Environmental Toxicology*, ed. K. Ranga Rao, Plenum Press, New York (1978), pp. 313–323.
46. Buser, H. R., and Rappe, C. Identification of substitution patterns in polychlorinated dibenzo-p-dioxins (PCDDs) by mass spectroscopy, *Chemosphere* **7**, 199–211 (1978).

2

Toxicology of Dioxins

Like all chlorinated dibenzo-*p*-dioxins, the tetrachlorinated isomer 2,3,7,8-tetrachlorodibenzo-*p*-dioxin (TCDD) is stable to heat, acids, and alkali. TCDD is virtually insoluble in water (2×10^{-4} ppm) only slightly soluble in fats (44 ppm in lard oil), and more soluble in hydrocarbons (570 ppm in benzene), and at its most soluble in chlorinated organic solvents (1400 ppm in ortho-dichlorobenzene).[1]

TCDD is almost immobile in soil. Even in very porous ground it remains close to the point of application,[2] where it is adsorbed on soil particles. There is little danger, therefore, of the dioxin leaching into the soil and contaminating ground water supplies.[3]

The dioxin is taken up from the soil by certain young plants, oats and soya beans in particular. However, no detectable amounts of the dioxin were found in the grain and soya beans harvested at maturity.[4] The authors of this report, Kearney *et al.* (1973) conclude, therefore, that even in the presence of significant amounts of TCDD in soil, uptake by plants is highly unlikely. Similarly they report that following foliar application of TCDD to the leaves of oat and soybean plants, no translocation occurred to other portions of these crops.

Dioxin Degradation

Microbial Action

Few microorganisms capable of degrading TCDD have been observed. Of 100 microbial strains examined, all of which we know to degrade persistent pesticides, only five exhibited any ability to detoxify TCDD. And in these strains the efficiency of dioxin degradation was low, with manipulation of the culture conditions resulting in no improvement in efficiency.[5] The rate of degradation of TCDD in soil samples would appear to be a function of the type of microorganisms present and the concentration of the dioxin. The half-lives of TCDD in soil from Utah and Florida were 320 and 230 days, respectively.[6] The dioxin half-life at Seveso is reported to be of the order of 2-3 years[7] and possibly longer (see Chapter 9).

Heat

Only 50% of TCDD is decomposed following exposure to a temperature of 700°C for 21 seconds. However, the same exposure at a temperature of 800°C will decompose the dioxin completely.[8]

Heat is also required for the production of TCDD. The chemical's synthesis in trichlorophenol manufacture is described in the section on the chemistry of the dioxin. But TCDD is also reported to be formed when the herbicide 2,4,5-T is heated. Pyrolysis of 2,4,5-T at temperatures of 500-600°C was reported by Buu Hoi et al.[9] to yield a pyrolysate which contained as much as 15% of TCDD. Pyrolysis of Silvex, a 2,4,5-T-based product, is said to also yield considerable quantities of TCDD[10].

However, the mass spectra data used to identify TCDD in the two preceding reports has been challenged. Langer et al.[11] failed in their attempt to repeat the observations of Buu Hoi and Saint Ruf.[10] Not only did Langer et al. fail to find TCDD but they suggested that the mass spectrum reported by Buu Hoi et al.[9] and Saint Ruf[10] was not that of TCDD but that of another polymeric substance. Langer et al.[11] concluded from their experiments that it was most unlikely that TCDD would be produced in the field by burning vegetation sprayed with 2,4,5-T.

It has been reported recently, however, that when vegetation sprayed with 2,4,5-T is burned, TCDD is indeed formed. Stehl and Lamparski,[12] analyzing the ashes after grass and paper had been sprayed with 2,4,5-T and subsequently burned concluded that some 0.6 ppt (parts per trillion) of TCDD would be produced for every 1 ppm of 2,4,5-T burned. Similar results but an order of magnitude higher have been reported by Ahling et al.[13] This group suggests that for every 1 ppm of 2,4,5-T burned at 500°C, 6 ppt of TCDD is formed.

Formation of TCDD by burning 2,4,5-T in small fires may be a problem, but it has been suggested that its formation in forest fires is not one.[14] Forest fires reach temperatures in excess of 1200°C. With TCDD known to decompose at a temperature above 800°C it is unlikely that much will be produced in a forest fire.[14]

Photochemical Decomposition

Chlorinated dibenzodioxins are decomposed by ultraviolet light providing the chemicals are first dissolved in suitable solvents. With methanol as solvent TCDD was first photolyzed using artificial ultraviolet light. Twenty-four hours was required for complete photolysis. Similar results were obtained when TCDD, in methanol, was exposed to natural sunlight; in this case 36 hours was required for photolysis. The photolytic process begins with the removal of chlorine atoms from the benzene ring followed by cleavage of one of the dioxin oxygen bridges.[15]

It has been reported that where TCDD is applied directly to soil or a solid surface the dioxin is extremely resistant to the action of sunlight and in consequence decomposition is slow.[15] For pure TCDD any loss of the chemical from the soil surface would occur by mechanical transfer such as the movement of dust particles rather than by volatilization or photodegradation.[15]

The requirements for photodegradation of TCDD are that the dioxin should be dissolved in a light-transmitting film, that an organic hydrogen donor should be present and that ultraviolet light be available.[16] Herbicides themselves can act as a solvent and hydrogen donor. Formulations of herbicides containing known amounts of TCDD and exposed to natural sunlight on leaves, soil, or grass are reported to lose most, and in some cases all, of the TCDD in a single day through photochemical dechlorination.[16] This would suggest that where 2,4,5-T spraying occurs environmental residues of TCDD might be less than was initially calculated.

In conditions where TCDD is known to be present on vegetation solubilizing agents can be applied to dissolve the chemical and make it available for photodegradation. One such experiment was carried out at Seveso in Italy. A solution of 80% olive oil and 20% cyclohexanone—to reduce viscosity—was sprayed onto vegetation. The results showed that this mixture enhanced photodegradation of TCDD.[7] Although fears had been expressed that application of the oil would increase the penetration of TCDD into the soil the results showed that this did not occur.[17]

Volatilization of TCDD is also reported to aid its photodegradation. Using a "microagroeco-system chamber" Nash and Beall[18] showed that when TCDD was volatilized, dechlorination, and hence detoxification, occurred in direct sun, sunlight filtered through glass, and even in outdoor shade. Thus they concluded that TCDD was even susceptible to photodegradation in the absence of ultraviolet light.

Chlorinolysis

In chlorinolysis, carbon compounds are mixed with chlorine at high temperatures and pressure. The process results in the conversion of the carbon compound to carbon tetrachloride. The herbicide formulation Agent Orange [a 1:1 mixture of the butyl esters of 2,4-dichlorophenoxyacetic acid (2,4-D) and 2,4,5-T] known to contain small amounts of TCDD has been subjected to chlorinolysis. No TCDD was detected in the carbon tetrachloride produced from the herbicide.[19]

Oxidation

The powerful oxidizing agent ruthenium tetroxide is reported to degrade TCDD efficiently.[20] Ruthenium tetroxide reacts with a wide range of organic substances and it can be used in solution in water or in organic solvents with no nucleophilic character such as chloroform, nitromethane, or carbon tetrachloride. Both 2,7-dichlorodibenzo-*p*-dioxin (DCDD) and TCDD are degraded by ruthenium tetroxide. A solution of ruthenium tetroxide in carbon tetrachloride will destroy DCDD if the mxiture is left overnight at ambient temperature. The rate of oxidation increases with increase in temperature: at 30°C DCDD has a half-life of 215 minutes, while at 50°C the half-life falls to 38 minutes. The oxidation of TCDD is a slower process: at 20°C the chemical has a half-life of 560 minutes in the ruthenium tetroxide mixture. At 70°C, however, the half life falls to 15 minutes. Only catalytic amounts of the ruthenium are required for the process after which it can be regenerated using a consumable secondary oxidant such as hypochlorite.[20]

Toxicological Properties

Distribution in Body and Metabolism

From the few reports available on the tissue distribution and excretion of TCDD it is evident that the chemical is rapidly but incompletely absorbed from the gut.[21-24] Because of its partition coefficient—the dioxin is more soluble in fats than aqueous solution—TCDD tends to be stored in fatty tissue. The liver is the principal storage site for TCDD in the rat[21] and guinea pig.[25]

Absorption of TCDD from the gut depends on the availability of the chemical. When dissolved in ethanol TCDD is easily absorbed from the intestine of rats.[26] However, when TCDD was mixed with soil particles, adsorption of the TCDD onto the particles occurred and absorption from the gut fell by half. Adsorption of TCDD to activated charcoal almost reduced intestinal uptake entirely. Similar results were observed when TCDD was applied to rat skin. Dermal absorption of TCDD was half that which occurred from the gut.[26] Adsorption of the chemical to soil particles or charcoal markedly reduced absorption through the skin as well.

Other vehicles for TCDD also affected its penetration through the skin. A lipophilic ointment decreased the penetration—the TCDD remaining dissolved in the ointment—whereas with a more hydrophilic ointment [polyethylene glycol/water (85:15:v/v)]—penetration was similar to that observed when TCDD dissolved in methanol was applied to the skin.[26]

Accumulation of TCDD in the liver appears to be a rapid process. Poiger and Schlatter[26] report that after a single oral dose of radioactive TCDD some 36% of the radioactivity was present in the liver within 24 hours. Two days later the value had fallen to 22% and a week after dosing stood at 17.5% of the original dose. Piper et al.,[21] on the other hand, using a TCDD concentration 700 times higher, recorded liver values of 47% and 45%, three and seven days, respectively, after oral dosing of the animals. The concentration of the chemical—perhaps causing more organ damage in the animals of Piper et al.[21]—and the strain of rat may account for these differences.

In adult male guinea pigs the distribution of TCDD in tissues is not the same as that observed in rats. Twenty-four hours after a single oral dose of ^{14}C-TCDD was fed to guinea pigs, the percentage distribution of radioactivity [% of original dose/g (gram) of tissue] was as follows: adipose tissue (2.36), adrenals (1.36), liver (1.13), spleen (0.70), intestine (0.42), and skin (0.48). The level of ^{14}C-TCDD in the liver increased threefold after a further two weeks and smaller increases were also noted in the adrenals, kidneys, and lungs. Only skin and adipose tissue recorded falls in the isotope.[25]

Other species where TCDD has been observed include fish, cattle, rhesus monkeys, and humans. Spotted sun fish taken from a 2,4,5-T sprayed area in Florida were recorded as having 4, 4, 18, and 85 ppt of TCDD in skin, muscle, gonads, and gut, respectively.[27] Ranges of TCDD from 20 to 60 ppt have been reported in a small percentage (3.5%) of samples of beef fat taken from cattle known to have been exposed to 2,4,5-T[28,29]. No detectable levels of TCDD were observed in the livers of these animals.[28]

Accumulation of TCDD in the rhesus monkey occurs in the skin, adipose tissue, and muscle after oral dosing.[30] As far as humans are concerned TCDD has been detected in human breast milk; levels of TCDD from 40 to 50 ppt were measured in samples collected (from women residing in areas in Vietnam heavily sprayed with the herbicide Agent Orange) in 1970 and analyzed 4 years later.[31] Samples of breast milk obtained from mothers residing in 2,4,5-T-sprayed areas in the United States had no detectable levels of TCDD—minimum detection level 10 ppt.[32] Levels of 3–57 ppt of TCDD have also been measured in fat tissue taken from 33 former Vietnam War veterans.[33] Curiously, measurable levels of dioxin were also detected in fat from ten individuals in the matching control group.[33]

The half-life of TCDD—administered as a single dose—is about 1 month in both the rat and guinea pig.[24,25] When TCDD is administered continuously it reaches a "steady state" level in the rat which appears to be independent of the dose.[24] At one stage it was thought that TCDD was not metabolized *in vivo*[21]; however, it is now known that there are at least two metabolites,[34] which the authors Poiger and Schlatter believe suggests that TCDD is excreted into bile in a metabolized form as a water-soluble conjugate. Conjugation probably occurs with phenolic hydroxyl groups, but their position on the TCDD molecule is not yet known.[35] The authors now believe that this metabolism is a means of detoxifying the chemical as the acute toxicity of one of the metabolites in the dog is at least 100 times lower than that of TCDD itself.[35]

Excretion of TCDD occurs principally via the feces, urine carrying only trace amounts of the chemical. The combined excretion of TCDD in guinea pig feces and urine appeared to be linear for the period of measurement—23 days with half of the administered chemical being excreted in about 30 days.[25] The rate of excretion is 4–6 times more rapid in the dog. In animals dosed with 0.1–0.2 mg (milligrams) of TCDD/kg (kilogram) body weight, some 20%–30% of the absorbed dose is excreted in the dog within 5 days.[35]

It appears that the rate of clearance of TCDD from the liver can be increased by the addition of either charcoal or cholic acid to the diet of mice. The amount of TCDD in the livers of mice fed either charcoal or cholic acid, or both, was well below that seen in animals not receiving these supplements. In one group the reduction was as much as 36%.[36] The supplements also appear to have an effect on the toxicity of TCDD. Sixty days after mice were given 90 mg TCDD/kg body weight no deaths were recorded in the 16 animals receiving charcoal or the 11 getting cholic acid supplements, whereas 45% of the respective control animals died within this period.[36]

General Toxicology

The toxicity of the chlorinated dibenzodioxins depends both on the number of chlorine atoms present and on their position on the parent dibenzodioxin ring structure. A comprehensive review of chlorinated dibenzodioxin toxicology is given in McConnell *et al.*,[37] from which the details in Table III are taken. It is clear from the table that 2,3,7,8-tetrachlorodibenzo-*p*-dioxin is the most toxic of the dioxin isomers. The molecule has four chlorine atoms, two each at extreme ends of the benzene rings. It is the number and positioning of these chlorines which gives this chemical its extreme toxicity. Removing just one of the chlorine atoms to form 2,3,7-trichlorodibenzo-*p*-dioxin reduces the toxicity by a factor of 15,000. Similarly, removing an additional chlorine atom reduces the toxicity even further; the LD_{50} value for 2,8-dichlorodibenzo-*p*-dioxin is shown in the table. (The LD_{50} value is the quantity of chemical, given in a single dose, required to kill half the test animals.)

TABLE III. Estimated Single Oral LD 50/30 Values of Chlorinated Dibenzo-p-Dioxins (McConnell et al.)[a]

Chlorination	Guinea pigs		Mice	
	μg/kg	μmol/kg	μg/kg	μmol/kg
2,8	>300,000	>1,180	—	—
2,3,7	29,444	120.41	>3,000	>10
2,3,7,8	2	0.006	283.7	0.88
1,2,3,7,8	3.1	0.009	337.5	0.94
1,2,4,7,8	1,125	3.15	>5,000	>14
1,2,3,4,7,8	72.5	0.185	825	2.11
1,2,3,6,7,8	70–100	0.178–0.255	1,250	3.19
1,2,3,7,8,9	60–100	0.153–0.255	>1,440	>3.67
1,2,3,4,6,7,8	>600	>1,400	—	—
1-NO$_2$-3,7,8	>30,000	>90	—	—
1-NH$_2$-3,7,8	>30,000	>99	—	—
1-NO$_2$-2,3,7,8	47.5	0.129	>2,000	>5.4
1-NH$_2$-2,3,7,8	194.2	0.576	>4,800	>14.2

[a] Reference 37.

Chlorination of 2,3,7,8-tetrachlorodibenzo-p-dioxin has a less marked effect on its toxicity than dechlorination. Addition of one chlorine atom to form 1,2,3,7,8-pentachlorodibenzo-p-dioxin reduces the LD$_{50}$ value, but only by a factor of 0.5. Further chlorination reduces the toxicity even more as shown by the LD$_{50}$ value for 1,2,3,4,7,8-hexachlorodibenzo-p-dioxin (see Table III). The structure of the hexachlorinated dioxin is shown in Figure 12.

The chlorinated dibenzofurans have a similar structure to the dibenzodioxins, the difference being that in the former compounds there is only one oxygen bridge, whereas in the latter compounds there are two. The toxicity of the dibenzofurans is similar to that of the dibenzodioxins. Toxicity studies show that the LD$_{50}$ value of 2,3,7,8-tetrachlorodibenzofuran is only two- to fourfold lower than that of 2,3,7,8-tetrachlorodibenzo-p-dioxin.[38]

Just as the toxicity of the chlorinated dibenzo-p-dioxins varies according to the degree of chlorination, so too does the toxicity of 2,3,7,8-tetrachlorodibenzo-p-dioxin vary in different animal species. Single toxic dose values for TCDD in various animal species are shown in Table IV. The guinea pig is the most vulnerable mammal with the lethal dose of TCDD as low as 0.6–2.0 μg/kg (micrograms per kilogram) body weight.[37,39] The LD$_{50}$ doses for rats are an order of magnitude higher than the dose for guinea pigs.[39,40] And the toxic dose is even higher in the other animal species tested, with the exception of the chicken.[45]

An equally striking feature of the results in Table IV is the time required for a single dose of TCDD to cause death in the animals. The time

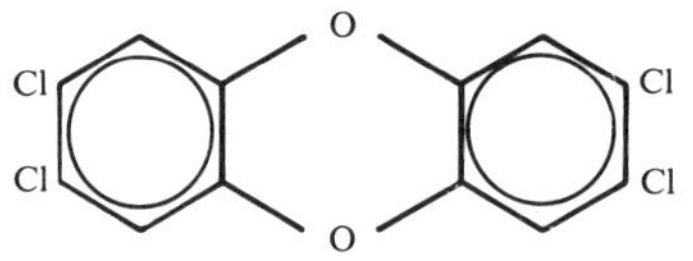

2,3,7,8-tetrachlorodibenzo-p-dioxin

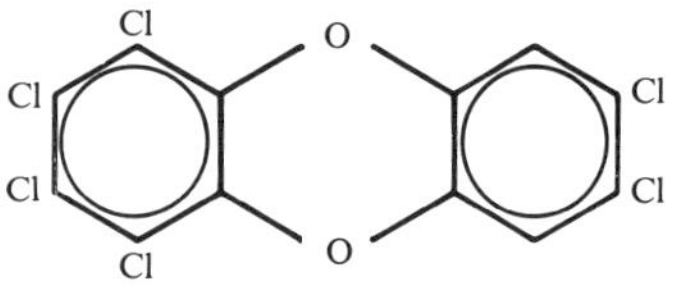

1,2,3,4,7,8-hexachlorodibenzo-p-dioxin

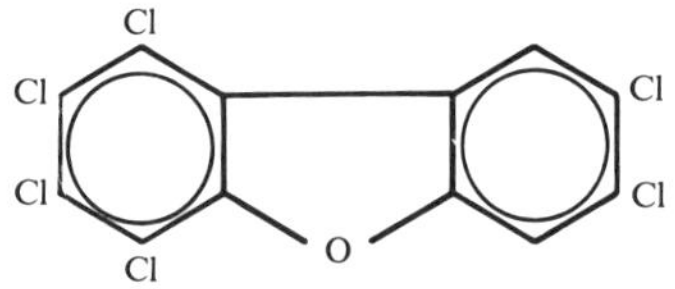

2,3,7,8-tetrachlorodibenzofuran

FIG. 12. Structures of two chlorinated dibenzo-*p*-dioxins and dibenzofuran.

of death can vary by as much as sixfold, as shown by the results for guinea pigs and rabbits.[39]

Pronounced weight loss and anorexia are common findings in most animals fed lethal concentrations of TCDD.[40,47–49] Atrophy of the thymus is a constant finding in all animals given a lethal dose of the chemical.[40,42,47,50,51] According to Vos *et al.*[52] sublethal doses of TCDD, 2–5 μg/kg body weight, affect the lymphoid system in the rat causing a suppression of cell-mediated immunity. The offspring of pregnant rats fed sublethal doses of TCDD on days 14 and 17 of gestation and postnatally on days 1, 8, and 15 had a lowered lymphocyte count in the thymus cortex. This fall resulted in the impairment of cellular immunity and an increase in the time taken to reject allografts.[52]

The depression in thymus-dependent immune function may explain the observation of Greig[53] that many rats poisoned with TCDD die with a severe lung infection. But as Greig points out, this in itself is not sufficient to explain the toxic action of TCDD since deaths have occurred in both TCDD-treated germ-free and SPF rats maintained in sterile conditions.[45] It is known that the resistance of rats to bacterial infections with *Salmonella* is reduced by TCDD,[54] but Vos[55] believes that this may be due to endotoxin

TABLE IV. Single Toxic Dose Values for 2,3,7,8-Tetrachlorodibenzo-p-Dioxin

Species	Strain	Sex	Route	LD50 (μg/kg)	Time to Death (days)	Reference
Guinea pig	Unknown	M	Oral	0.6	5-34	Schwetz et al.[39]
	Hartley	M	Oral	2	17-20	McConnell et al.[37]
Rat	Spartan	M	Oral	22	9-27	Schwetz et al.[39]
	Spartan	F	Oral	45	13-43	Schwetz et al.[39]
	CD	M	Oral	<100	18	Harris et al.[40]
	CD	F	Oral	>50 <100	18	Harris et al.[40]
Monkey	Rhesus	F	Oral	<70	28-47	McConnell et al.[41]
	Rhesus	M	Oral	<1	12-78	McNulty[42]
Mouse	C57 BL/6	M	Oral	114	20	Vos et al.[43]
	C57 BL/6	M	Oral	126	21	Jones and Greig[44]
	C57 BL/6	M	Oral	284	22-25	McConnell et al.[37]
Rabbit	Unknown	Mixed	Oral	115	6-39	Schwetz et al.[39]
	Unknown	Mixed	Skin	275	12-22	Schwetz et al.[39]
	Unknown	Mixed	I.P.	>63 <500	6-23	Schwetz et al.[39]
Dog	Unknown	M	Oral	>30 <300	9-15	Schwetz et al.[39]
	Unknown	F	Oral	>100	—	Schwetz et al.[39]
Chicken	Leghorn	Unknown	Oral	25-50	12-21	Greig et al.[45]
Hampster	Golden	M/F	I.P.	>3000	1->50	Olson et al.[46]
	Syrian	M/F	Oral	1157	2-47	

present in the *Salmonella*—TCDD-treated mice show a marked increase in susceptibility to endotoxin.

The immunosuppressive effect of TCDD is not restricted to rats which are considered to be poor immunologic responders; the effect occurs equally in rats which are good immunologic responders.[56] Although TCDD's mechanism of induction of immunosuppression is unknown it is clear that the effect is reversible, at least in mice which are good immunologic responders.[56]

The only available evidence on TCDD's immunosuppressive effect in humans is in a report—unpublished—describing an investigation of workers all of whom had chloracne and who were exposed to the chemical between 1968 and 1971. According to the investigator, Ward,[57] there was no significant difference between the TCDD-exposed group and a control group with respect to total lymphocyte count, T-cells, or B cells. The exposed group, however, had a higher proportion of cases with reduced levels of the immunoglobulins IgD and IgM. Furthermore, the exposed group had more cases in which the lymphocyte transformation in response to phytohemaglutin was reduced. The author suggests that the TCDD-exposed workers are relatively deficient in primary immune capability and in T-cell–B-cell cooperation, and he suggests that toxic exposure to TCDD induces a lasting effect on B-cell memory and immune capability.[57] (For a fuller discussion of these findings refer to the section on Coalite in Chapter 6.)

Hepatoxicity

The liver is a major target organ for TCDD and is severely affected by the chemical. TCDD will cause extensive necrosis of the liver in rabbits,[58,59] but a more localized focal centrolobular necrosis in rats.[47,50,53,60,61] The size and shape of hepatocytes in rats show considerable variations, and large multinucleated hepatocytes and nuclear enlargements have also been observed.[42,60] Electron microscope studies have shown that the multinucleate cells arise by fusion of parenchymal cells.[61] According to Greig[53] the formation of those abnormal cells appears to be the result of changes at the cell membrane soon after dosing. Histochemical studies show that ATPase activity disappears rapidly,[62,63] indicating that the parenchymal cell membrane is a target site for the toxic action of TCDD. The loss of ATPase is also reported to have been observed in biochemical studies of isolated membrane preparations.[53]

These membrane changes may be related to a disordered lipid metabolism.[64] TCDD induces marked enlargement of the liver in rats and mice, and lipid accumulation is marked.[43,44,50] The dioxin is reported not to affect the livers ability to synthesize lipids[48] as it has no effect on the incorporation of ^{3}H-acetate in lipids; however, as lipid does accumulate in liver, it seems that its transport out of this organ is affected by the chemical.

Hepatic porphyria is easily produced in mice and a 4000-fold increase in uroporphyrin has been reported,[65] the result, perhaps, of inhibition of the enzyme uroporphyrinogen decarboxylase.[66] Prolonged dosing of TCDD will cause porphyria in the rat,[67] and in the chick embryo the chemical stimulates the rate-limiting enzyme of porphyrin biosynthesis, δ-ALA synthetase.[68] But TCDD does not produce porphyria in guinea pigs[60] or primates.[69] Conflicting results have been reported for humans. Laboratory workers exposed to TCDD are reported not to have developed porphyria,[70] whereas other workers exposed to the chemical while handling 2,4,5-trichlorophenol are said to have developed the condition.[71,72]

One of the main toxic effects of TCDD in the rat and rabbit is damage to the liver.[55] The hepatotoxic effects observed in the rat include elevations of the serum hepatic enzymes glutamic oxaloacetic transaminase (SGOT) and glutamic pyruvic transaminase (SGPT), as well as hyperbilirubinemia, hypercholesterolemia, and hypoproteinemia.[73] Similar effects have been reported in humans. Workers exposed to TCDD have raised serum triglyceride and cholesterol[70,72,74] values and elevated levels of SGOT and SGPT.[72,75]

There is a marked species variation in the hepatotoxic action of TCDD which may be due in part to the presence or absence of receptors in the liver with a high affinity for the chemical.[55] TCDD induces the liver enzyme aryl hydrocarbon hydroxylase, and the induction is most marked in species with receptors having a high affinity for TCDD compared with those with low-affinity receptors.[55] The former group, termed *responsive mice*, have a greater hepatic uptake of TCDD than the latter, *unresponsive*, animals. It has been mooted that a hepatoxic effect may be mild in species which have low-affinity receptors making them poor responders.[55]

Edema

Water retention, or edema, is considered to be the most characteristic symptom of TCDD poisoning in chickens. Fluid accumulation occurs in the hydropericardium heart sac, peritoneal cavity, in the lung, and in subcutaneous tissue.[39]

An outbreak of "chick edema disease" occurred in the Midwest in the United States in 1957, with millions of broiler chickens dying.[76] Toxic animal fat fed to the chickens as feed stock was rapidly isolated as the source of the problem. The feed stock had been prepared from hides previously treated with a preservative used in the tanning industry, pentachlorophenol. The pentachlorophenol was contaminated with 1,2,3,7,8,9-hexachlorodibenzo-*p*-dioxin and it was the dioxin which was identified as causing the edema.[77] Other dioxins including TCDD have also been identified in the fat.[78]

Mammalian species exposed to TCDD also develop edema. Rat fetuses treated prenatally with the chemical developed subcutaneous edema in the head, neck, and trunk.[79] Severe terminal edema has been observed in about a quarter of mice receiving a lethal dose of the chemical.[44,52] The edema occurred in subcutaneous tissue and in abdominal and thoraccic cavities. Primates, too, developed edema after TCDD intoxication. The edema, noticeable in the lips, is accompanied by reduced serum albumin levels.[69,80]

Less Specific Toxic Effects

TCDD causes a variety of other lesions in different animal species. Damage to kidney tubular epithelial cells occurs in rats given lethal doses of the chemical.[60] Ulceration and necrosis of the glandular section of the stomach,[69] chloracne, loss of eyelids, facial alopecia, and abnormal growth and loss of toe and finger nails also occur in monkeys exposed to TCDD.[80] Testicular atrophy has been described in both primates[69] and rats.[50]

Horses exposed to TCDD experienced loss of hair, chronic emaciation, skin lesions, edema, intestinal colic, conjunctivitis, and joint stiffness.[81] These findings are described in more detail in the section on the Missouri accident in Chapter 6.

Fish are extremely sensitive to TCDD. Low doses of the chemical are extremely toxic to fish. The first symptom of toxicity is a declining interest in food. This is followed in guppies (*Paecilia reticulatus*) Peters by skin discoloration and erosion of the upper jaw.[82,83] Depressed body growth and food consumption occurs in young rainbow trout (*Salmo gairdineri*) fed a daily TCDD ration [2.3 ng/kg (nanograms per kilogram) dry feed]. Fin necrosis occurred 14 days after commencement of the diet, with death occurring after a further 19 days.[82]

Clinical Effects

Although the clinical symptoms of TCDD exposure are discussed in detail in other chapters, it is perhaps useful to summarize them in a section on toxicology. There have been 24 recorded accidents in chemical plants manufacturing trichloro and pentachlorophenol, and the clinical information below is collated from reports on these incidents.[7,57-59,70-72,74,75,84-95] These symptoms include the following.

  a. Skin Changes: Chloracne, hyperpigmentation, hirsutism.

  b. Systemic Effects: Liver damage (mild fibrosis); raised serum hepatic enzymes, glutamic oxaloacetic transaminase (SGOT), and glutamic pyruvic transaminase (SGPT); increased excretion of porphyrins in urine; disorders of fat metabolism (hypertriglyceridemia, hypercholesterolemia) and carbohydrate metabolism; cardiovascular, urinary tract, respiratory, pancreatic, and digestive disorders (flatulence, nausea, vomiting, diarrhea); loss of appetite and weight loss; muscular aches and pains, and pain in joints; reduced primary immune capability.

  c. Neurological Effects: Polyneuropathies; lower extremity weakness; impairment of sensory functions including sight disturbances, loss of hearing, taste, and sense of smell; headaches.

  d. Psychiatric Effects: Depression, loss of energy and drive, disturbance of sleep, uncharacteristic bouts of anger.

Teratogenic Effects

A chemical is a teratogen when it causes developmental disturbances in the embryo which result in congenital malformation. If a chemical kills the embryo it is said to be *embryocidal*, and if it produces tissue damage (not necessarily resulting in malformation) it is *embryopathic*. The term *embryotoxic* generally refers to any harmful effect on the embryo.

Most of the teratogenic and embryotoxic studies involving TCDD have been conducted using the herbicide 2,4,5-T.[96] Various formulations of the herbicide with a TCDD content of 0.02–30 ppm have been shown to be teratogenic for mice.[97,98]

TCDD causes terata in the C 57 Bl/6 strain of mice at a single dose of 1 µg/kg when given on day 10 of gestation.[101] In other mice strains doses of 1–3 µg/kg administered on days 6–15 of gestation will cause kidney abnormalities and cleft palate in most animals. (See Table V for details.) It is also clear from the results available that there is a difference in sensitivity to TCDD between strains of mice. For example, 3 µg/kg of TCDD on days 6–15 of gestation is required to produce cleft palates in the NMRI mouse strain,[96] whereas 1 µg/kg over the same period will have this effect in the CD-1 strain.[100]

The proportion of fetuses born malformed increases as the TCDD dose increases as shown by the differences between the dose which will produce a teratogenic effect and the corresponding ED_{50} dose (the concentration required to affect 50% of the animals).[96–98] Repeated daily oral doses of TCDD from 25–400 µg/kg had an increasing fetotoxic and teratogenic effect in mice with some 97% of the animals affected at the highest dose.[99]

2,4,5-T itself is a teratogen in animals at dose levels 10,000 times that of dioxin. In some strains of mice the herbicide is a teratogen at a dose of 35 mg/kg administered on days 6–15 of pregnancy.[103] The TCDD content of the 2,4,5-T tested was 0.05 ppm. Where a concentration of 35 mg 2,4,5-T/kg is fed to animals the TCDD concentration would be approximately 2 ng/kg and thus 1000 times lower than the concentration required to produce a teratogenic effect. In many cases, therefore, 2,4,5-T presents a teratogenic risk before the TCDD content of 2,4,5-T does. Concentrations of 2,4,5-T from 30–150 mg/kg administered on days 6–15 of pregnancy are teratogenic in mice with cleft palate and kidney abnormalities the most common findings.[96,104] The higher the dose of 2,4,5-T, the greater the teratogenic effect. In general, the TCDD content of the 2,4,5-T tested has been less than 0.1 ppm. In some cases, however, it has been as high as 1.5 ppm. At this concentration, however, cleft palates occurred in mice given 30–60 mg/kg on days 6–15 of pregnancy.[96]

Only small amounts of TCDD are required to produce fetal abnormalities in mice and the incidence of malformations depends on the frequency of dosing. Repeated doses of 5 µg/kg of TCDD on days 7–11 of pregnancy produced cleft palate in 65% of fetuses, whereas 25 µg/kg given as a single dose on day 7 caused the condition in only 16% of fetuses. If a single dose of 25 µg/kg TCDD is given on day 10, however, 84% of fetuses developed cleft palate.[105] Only 1.4% of the control fetuses had this malformation.

The concentration of TCDD found in the embryo and fetus of pregnant mice administered TCDD is very low between gestational days 11 and 18. TCDD concentrations in the fetus were between 0.04% and 0.14% of

TABLE V. Teratogenic and Embryotoxic Effects of TCDD

Species	Strain	Effect	Minimum dose used (μg/kg)	Duration of administration (Number of days of gestation)	Route/frequency	Reference
Mouse	NMRI	Cleft palate	3	6–15	Oral/daily	96
		Cleft palate	6.5, ED_{50}	6–15	Oral/daily	96
	NMRI	Cleft palate	9	9–13	Oral/daily	96
		Cleft palate	9, ED_{50}	9–13	Oral/daily	96
		Cleft palate	15^a	13	Oral/single dose	97
			40, ED_{50}	13	Oral/single dose	97
		Cleft palate	5^a	11	Oral/single dose	97
			15, ED_{50}	11	Oral/single dose	97
	CD-1	Cleft palate	1^a	6–15	Subcutaneous/daily	98
			>3, ED_{50}	6–15	Subcutaneous/daily	98
		Kidney abnormality	1	6–15	Subcutaneous/daily	98
		Kidney abnormality	1–3, ED_{50}	6–15	Subcutaneous/daily	98
	DBA/2J	Cleft palate	3^a	6–15	Subcutaneous/daily	98
		Kidney abnormality	3^a	6–15	Subcutaneous/daily	98
	C57B1/6J	Cleft palate	3^a	6–15	Subcutaneous/daily	98
		Kidney abnormality	3^a	6–15	Subcutaneous/daily	98
	CD1 31 pregnant females	Fetotoxic, teratogenic, increasing effect with dosage—97% at higher dose	25, 50, 100, 200, 400	7–16	Oral/daily	99
		Fetotoxic, teratogenic, increasing effect with dosage—76% at highest dose	25, 50, 100, and 200	7–16	Subcutaneous/daily	99
	CD1 17 pregnant females	No effect	0.1	6–15	Oral/gavage daily	100

	Strain	Effect	Dose	Days	Route	Ref
	CD1	Increased fetal resorption sites, 21% cleft palate, 5% kidney abnormalities	1	6–15	Oral/gavage daily	100
		71% cleft palate	3	6–15	Oral/gavage daily	100
		28% kidney abnormalities				
	C57Bl/6	34% kidney abnormalities	1	10	Oral/single dose	101
		1.9% cleft palate	1	10–13	Oral/daily	101
		58.9% kidney abnormalities				
		55.4% cleft palate	3	10–13	Oral/daily	101
		95.1% kidney abnormalities				
	C57Bl/6 fetuses	Renal hydronephrosis in 12%, 71%, or 75% depending on dose	1, 3, or 10	At parturition	Oral/females received single dose	101
Rat	Sprague-Dawley (Spartan)	No effect	0.03	6–15	Oral/daily	102
		Depressed fetal weight	0.125 and 2	6–15	Oral/daily	102
		Internal hemorrhages in fetuses	0.125, 0.5, or 2	6–15	Oral/daily	102
		All fetuses died	8	6–15	Oral/daily	102
Rat	Sprague-Dawley	No effect	0.125	6–15	Oral/daily	79
		Slight decrease in fetal weight	0.25	6–15	Oral/daily	79
		Reduced fetal survival, lower body weight, and lowered reproductive rate in pregnancy	0.5 and 1	6–15	Oral/daily	79
		Visceral lesions, reduced fetal weight, increased fetal death with maternal toxicity	1 and 2	6–15	Oral/daily	79
		100% embryo death	4	6–15	Oral/daily	79
	CD	No effect on fetal mortality	0.5	6–15	Subcutaneous/daily	98
		67% kidney abnormalities	0.5	6–15	Subcutaneous/daily	98

[a] This represents the smallest dose required to produce a significant teratogenic effect. In some instances, only one dose level was tested and this may not be the smallest quantity which will produce a teratogenic effect.

the maternal dose, two orders of magnitude lower than the concentration in maternal liver—the organ in which TCDD accumulation was greatest.[105]

The minimum dose of 2,4,5-T which is teratogenic in the rat appears to be 50 mg/kg administered orally on days 6-15 of pregnancy.[105,106] Doses of 100 mg/kg given orally on days 1-14 of gestation are reported to cause many fetal deformities, and doses as high as 400 mg/kg cause many limb abnormalities.[107]

Purified 2,4,5-T (2,3,7,8-TCDD content <0.03 ppb) has no effect on reproduction in rats at 3 mg/kg/day but at 30 mg/kg/day it is reported that there is some evidence of a reduction in the postnatal survival rate of litters.[108] The higher of the two doses was sufficient to cause signs of toxicity in weanling animals but had no effect on reproductive capacity other than on litter survival rates.[108] In the Golden Hamster doses of 2,4,5-T with TCDD concentrations of 0.5-45 ppm are teratogenic when given orally on days 6-10 of gestation. Doses of less than 100 mg/kg of pure 2,4,5-T have no such effect.[109]

2,4,5-T given in doses of up 40 mg/kg at regular intervals during pregnancy are not teratogenic in the rabbit or monkey,[110,111] and doses as high as 113 mg/kg have no teratogenic effect in sheep.[112]

There is no convincing evidence that formulations of 2,4,5-T used commercially in agriculture and silviculture have caused malformations in humans. An abnormally high incidence of neural tube defects (such as spina bifida) in three areas of North Island, New Zealand, was attributed to 2,4,5-T spraying in the area. Investigations by the New Zealand Department of Health in 1977 refuted this connection. The Health Department claims that in two areas the finding was a chance occurrence and that if there was a causal factor in the third area it was not 2,4,5-T.[113] A similar investigation in Australia of a high incidence of birth defects in the Yarram district of Victoria in 1975/1976 concluded that the abnormalities were not caused by exposure to 2,4,5-T or 2,4-D.[114] Burning of sugar cane stubble in Queensland, Australia, was alleged to have led to an increase in birth defects; however, this is refuted by the Australian National Health and Medical Research Council.[115]

Agent Orange, the 1:1 mixture of 2,4,5-T and 2,4-D used as a defoliant in Vietnam, is said to have caused child deaths among the Montagnard population in South Vietnam.[116] Vietnamese[117] and American scientists[118] have reported that the use of Agent Orange—known to have had a high TCDD content (0.05-45 ppm)—is associated with an increase in the incidence of cleft palate and spina bifida in children born to mothers resident in areas of South Vietnam which were heavily sprayed. However, these findings have been criticized. Owing to the lack of good hospital records for the areas concerned for a period prior to the onset of spraying, it has proved difficult to accept these findings as being conclusive.[116,119]

More recently, the U.S. Environmental Protection Agency has claimed that the use of 2,4,5-T is associated with an increase in the incidence of spontaneous abortions in the State of Oregon.[120] As a result of its findings the Agency imposed a temporary ban on the use of 2,4,5-T in forestry and rights of way.[121] The EPA's study and its conclusions have been severely criticized. Critics claim that the areas chosen by the EPA were improperly matched and that the statistical approach used to analyze the study's findings was wrong. Furthermore the critics claim that if 2,4,5-T sprayed areas are properly matched with a suitable control area and differences in hospital admission rates for miscarriages between areas are taken into account, there is no longer a link between 2,4,5-T spraying and an increased incidence of spontaneous abortions.[122]

Similar negative findings are reported by the Dow Chemical Company in a survey to assess the outcome of pregnancy in the wives of Dow employees who might have been exposed to TCDD. Three hundred and seventy wives were included in this group; the control group consisted of 345 wives whose husbands were unlikely to have been exposed to TCDD. No statistically significant differences were reported between the two groups for the occurrence of miscarriages, still-births, infant deaths, or congenital malformations.[123]

Mutagenicity

There is a widespread belief among cancer workers that DNA damage is involved in the induction of cancer.[124] A chemical which will damage DNA and cause a cell to mutate is classified as a mutagen. Assay systems, commonly employing bacterial, fungal, or mammalian cell lines, have been developed to detect mutagens. Chemicals which are carcinogens in animals are usually identified as mutagens by these assays, usually referred to as short term tests.[124-126] However, not every mutagen will cause cancer in animals. The fact that most mutagens probably do cause cancer has enabled the tests to be used to identify potential carcinogens. A recent international study has confirmed that there are short-term tests that can be used to predict carcinogenic activity.[127] All tests give false negatives and false positives too, and it is well known that tests give better results for one class of chemical than another. However, it is possible to select tests which are complementary and to include these in a battery of tests which will identify most potential carcinogens.[127]

Most chemicals requre metabolic activation before they are mutagenic. In mammals this activation is performed by microsomal enzymes in the liver. Most tests now employ, therefore, a liver microsomal preparation, the so-called S-9 mix, as an obligatory activation step.[128]

TCDD has been tested for mutagenicity using the *in vitro* short-term tests in both the presence and absence of the S-9 mix. Two groups have reported that when TCDD was tested in the Ames bacterial test using the *Salmonella typhimurium* strain TA1532 without the S-9 mix, a positive result was obtained, suggesting that the chemical was indeed mutagenic.[129,130] TCDD is also reported to be mutagenic in the *Escherichia coli* strain Sd-4.[129] The TA1532 strain which measures the reversion of the cell line from a dependence on histidine to independence (the strains are placed on medium without histidine and hence will only grow if they mutate and become independent of histidine, the factor limiting growth) is well known as a strain which detects frameshift mutagens and hence chemicals which cause mutagens by intercalating with DNA.[131] The chemicals insert themselves into DNA, between the base pairs, thus causing a partial unwinding of the DNA double helix. This effect might lead to a misreading of the genetic code during replication of the cell, and thus to mutation.

A positive result for TCDD with strain TA1532 has not been obtained in every laboratory, however. McCann[132] has used TA1532 as well as strains TA1535, TA1537, and TA1538 with and without the S-9 mix, but without success; negative results were obtained with all the strains. TA1537 and TA1538 are sensitive indicators of frameshift mutagens and have replaced strain TA1532 in most laboratories. It is all the more surprising therefore, that if TCDD is an intercalating agent it fails to give a positive result with the strains optimised to detect this. Other workers also report negative results for TCDD tested with strains TA1535 and TA1538.[133] Recent unpublished work also confirms that the chemical is negative with strain TA1537.[134]

Thus the evidence implicating TCDD as a mutagen is conflicting. However, in view of the well-known fickleness of short-term tests and their ability to detect certain classes of chemicals as mutagens, but failure to detect other types of chemicals,[127] testing of TCDD in another test system was required to resolve the confusing evidence of the dioxin's mutagenic properties. Such a study has been performed using a mammalian *in vitro* test, the BHK cell transformation assay.[135] In this test, which employs the use of baby hampster kidney cells, TCDD is undoubtedly positive.[134] (A pure and an impure preparation of TCDD were tested in this assay; both were mutagenic, a finding which indicates that the impurity had no effect on the test result.) The unsubstituted dibenzo-*p*-dioxin, a noncarcinogen in animals,[137] used as the structurally appropriate negative control chemical was negative in this assay,[134] a finding also observed with various *Salmonella* strains[136] (see Table VI). The dioxin isomer 2,8-dichlorodibenzo-*p*-dioxin was weakly positive in the BHK test, whereas octachlorodibenzo-*p*-dioxin was negative[134] (see Figure 13a–c).

The positive response obtained with the *Salmonella* strain TA1532 led the authors to conclude that TCDD was indeed mutagenic and that it inter-

Toxicology of Dioxins

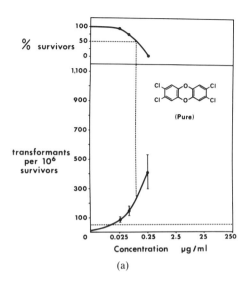

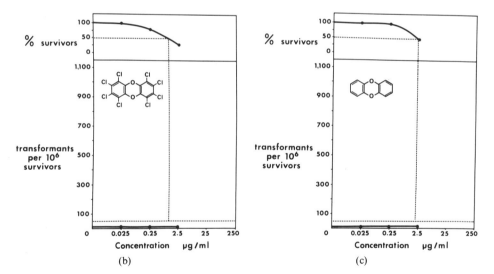

FIG. 13. Responses observed for TCDD, OCDD, and dibenzodioxin in the BHK cell transformation assay[135] are shown in parts (a), (b), and (c), respectively. The sample of TCDD was provided by Professor T. Inch and the remaining two chemicals were commercially available material. The data shown are the average of two experiments, a positive result being scored when the number of induced transformants per 10^6 cells exceeds five times the background incidence at the LD_{50} dose level. This subclone of BHK 21 C13 had a spontaneous transformation rate of 10 per 10^6 (unpublished results, Ashby *et al.*, 1980[134]).

TABLE VI. Mutagenicity of Dioxin Compounds in *Salmonella Typhimurium* and Baby Hamster Kidney Cells (BHK)[a]

Dioxin isomer[a]	Carcinogens[c] in animal studies	Strains detecting base pair substitutions				Strains detecting frameshifts					BHK	Reference
		G46	TA1530	TA1535	TA, 100	TA1531	TA1532	TA1534	TA1537	TA1538		
(i) impure	+	0	0	0	0	0	0	0	−	0	+	134
(ii) pure	+	0	0	0	0	0	0	0	0	0	+	134
		0	0	−	0	0	−	0	−	−	0	132
		0	0	−	0	0	0	0	0	0	0	133
		0	−	0	0	??	+	??	0	0	0	129
		−	0	0	0	0	+	0	0	0	0	130
(iii)	?	0	0	0	0	0	0	0	0	0	+	134

(iv) [tetrachlorodibenzo-p-dioxin structure]	?	0 / −	0 / 0	0 / 0	0 / −	0 / ??	0 / ??	0 / 0	− / 0	134 / 130
(v) [trichlorodibenzo-p-dioxin structure]	?	0	0	0	0	0	0	0	+	134
(vi) [dibenzo-p-dioxin structure]	−	0 / 0	0 / 0	0 / −	0 / 0	0 / 0	0 / −	0 / −	− / 0	134 / 136

[a] Adapted from data in Wasson et al.[131]
[b] (i) 2,3,7,8-Tetrachlorodibenzo-p-dioxin [95(%) pure]; (ii) 2,3,7,8-tetrachlorodibenzo-p-dioxin (pure)—it is assumed that unless stated tests were performed with a pure sample, although no evidence is available to confirm this; (iii) 2,8-dichlorodibenzo-p-dioxin; (iv) octachlorodibenzo-p-dioxin; (v) 1,3,7-trichlorodibenzo-p-dioxin; (vi) dibenzo-p-dioxin.
[c] +, positive carcinogen; −, noncarcinogen; ?, not known.
[d] +, positive; −, negative; 0, not tested; ??, doubtful mutagen.

calated with DNA. Where intercalation occurs the properties of both the chemical and the DNA are altered in a standard way. For example, the melting temperature of DNA is increased by 5–10°C, and the visible spectrum of the chemical is altered, undergoing a bathochromic (red) shift. In the case of the intercalating agent, Ellipticine, following intercalation the rise in the DNA melting temperature was 8.8°C, and the bathochromic shift about 40 mm.[138] When TCDD was tested by Ashby et al.[134] as a potential intercalating agent with calf thymus DNA it had no effect on the melting temperature, and no spectral changes were observed. On the basis of these findings the authors suggest that TCDD is not an intercalating agent. However, their evidence from the BHK test does confirm that it transforms cells *in vitro*. Just how it causes this effect has yet to be established.

Investigations of the cytological effects of TCDD indicate that it affects cell division. The chemical has a powerful inhibitory effect on mitosis in the African Blood Lily (*Haemanthus Katherinae*, Baker).[139] Other chromosomal effects in this plant included the formation of dicentric bridges and chromatin fusion with the formation of multinuclei or a single large nucleus.[139] Multinucleated cells have also been observed in mammals following treatment or poisoning with TCDD.[43,45,47,140]

In addition, the chemical is said to cause an increase in the number of chromosome aberrations in rat bone marrow in animals receiving twice weekly TCDD doses of 0.25, 0.5, 1.0, 2.0, or 4.0 μg/kg by gavage (stomach tube) for 13 weeks. The increase in chromosome breaks in both male and female animals receiving 4.0 μg/kg were significantly higher than those given 0.25 μg/kg.[141] However, these results are said to only indicate a weakly positive chromosome damaging effect of TCDD.[141]

The only evidence of chromosome damage in humans arising from exposure to TCDD is from Vietnam. Vietnamese scientists claim that there is a higher incidence of chromosome breaks and gaps in individuals exposed to the herbicide Agent Orange when it was used as a defoliant in Vietnam.[117] However, the level of chromosomal damage reported as abnormal by Tung et al.[117] is considered to be normal in the West. The results obtained from these investigations are based on a small number of samples and hence inadequate for statistical analysis.[119] A controlled epidemiological study with adequate numbers was not possible according to one of the authors as the study was performed at the height of the Vietnam War.[119]

No chromosomal abnormalities have been reported in workers engaged in the manufacture of 2,4,5-T at Dow Chemicals in the United States,[142] or in men exposed to TCDD as a result of industrial accidents in West Germany[143] or Britain.[144]

The fact that TCDD has a weak chromosome-breaking effect in rat bone marrow cells,[141] and that the chemical reduces spermatogenesis in

Toxicology of Dioxins

rats,[145] and testicular DNA synthesis in mice,[146] suggests that DNA in germ cells may be affected by the chemical. Direct effects on fertility have also been observed. Male and female rats in a three-generation study fed TCDD at 0.01 µg/kg had a lower survival rate, reduced growth rate and smaller litter size.[147] Primates fed TCDD at the parts-per-trillion level have irregular menstrual cycles, excessive hemorrhaging during menstruation, a reduced conception rate, and high incidence of early abortions.[148]

Demonstrating that TCDD has an effect on cells appears to be an easier process if the tests are performed *in vivo* than if they are done *in vitro*. The toxic effect of TCDD has been investigated on 23 cultured cell types but without success.[149] The cells were taken from tissues and/or species which are known to be susceptible to TCDD toxicity and included primary liver cells, cells from hepatomas, lymphomas, fibroblasts, and urinary epithelium. Morphology, percentage viability, and growth rate of the cell line were unaffected by TCDD. The authors[149] suggest that their failure to observe an *in vitro* toxic effect may be due to their not using a specific target cell for TCDD or perhaps because toxicity may require a cell–cell interaction of two different cell types.

Carcinogenicity

There is no doubt about the carcinogenicity of TCDD in animals. Several reports have shown unequivocally that this chemical will cause cancer in rats and mice (see Table VII for summary). The liver, lung, and thyroid appear to be the organs in which tumor development is most common.

In a feeding study performed by scientists at the Dow Chemical Company a concentration of 0.1 µg TCDD/kg body weight (2.2 µg TCDD/kg diet) caused an increased incidence of hepatocellular carcinomas and squamous cell carcinomas of the lung, hard palate/nasal turbinates, or tongue, whereas a reduced incidence of tumors was noted in the pituitary, uterus, pancreas, and mammary and adrenal glands.[151] Liver tumors and squamous cell tumors of the lung were also noted by Allen *et al.*[150] in rats fed a diet containing 1 and 5 µg TCDD/kg diet. In a third study rats and mice administered TCDD by gavage had a significant increase of tumors at a twice weekly dose of 0.5 µg/kg body weight.[152]

A dose-related increase—between the lowest and highest doses (see Table VII),—in thyroid and liver tumors was noted in female animals, whereas in males only a dose-related increase in liver tumors was observed.[152]

Only one of the three studies reports a carcinogenic effect for TCDD at a concentration of 5-mg/kg diet[150]—some 400 times lower than the concentration reported to be carcinogenic by Dow Chemical scientists. Allen *et al.*[150] report six tumors in five animals on this dose. They include an

TABLE VII. Review of Carcinogenicity Studies on Dibenzo Dioxins in Animals

Chemical	Animal	Species and no. used	Route of administration	Dose	Effect	Ref.
2,3,7,8-Tetrachloro-dibenzo-p-dioxin	Rat	Sprague-Dawley 10 males per group	Diet—for 78 weeks	Control 1 ng/kg diet 5 ng/kg diet 50 ng/kg diet 500 ng/kg diet 1 and 5 µg/kg diet	60% dead at 95th week — No tumors 20% dead at 95th week 40% dead at 95th week 40% dead at 95th week 50% dead at 95th week All died in 30–90 weeks — 38% with a range of tumor types. No common organ affected in three lower groups—lung and liver affected at 1 and 5 µg/kg	150
		Sprague-Dawley		50, 500 and 1,000 µg/kg diet	All died in 2–4 weeks	
		50 males and females per group		.001 µg/kg body wt (≡22 ng/kg diet) .01 µg/kg body wt (≡210 ng/kg diet)	No effect Nodules in liver — No carcinogenic response Hyperplesia in lung	151
				.1 µg/kg body wt (=2.2 µg/kg diet)	Increased incidents of liver, lung, hard palate and tongue tumors—reduced incidents of pituitary, uterine, mammary gland, pancreas, and adrenal gland tumors.	
		Osborne-Mendel 50 males and females per group	Feeding directly into stomach (by gavage) 2 days/week for 104 weeks	0.01, 0.05 and 0.5 µg/kg/week.	Dose-related increase in incidence of follicular-cell thyroid tumors in males. Dose related increase in incidence of follicular-cell thyroid tumors and liver tumors in females.	152
	Mice	B6C3F1 50 males and females per group	By gavage 2 days/week for 104 weeks	Males—0.01, 0.05, and 0.5 µg/kg/week	Dose-related increase in incidence of liver tumors	152

Compound	Species	Strain/Group	Route	Dose	Effect	Ref
		Swiss-Webster 30 males and females per group	Dermal application in acetone suspension 3 days/week for 104 weeks	Females 0.04, 0.2, and 2 µg/kg/week	Dose-related increase in thyroid and liver tumors	153
				Males—0.001 µg per application Females—0.005 µg per application	No effect. Increased incidence of fibrosarcoma in skin	154
		Swiss/H/Riop 45 males per group	By gavage (weekly)	0.007 µg/kg body weight 0.7 µg/kg body weight 7 µg/kg body weight	No increase in tumors. Significant increase in liver tumors. Life span of animals reduced a third. No marked increase in tumors observed.	
	Monkey	*Macaca mulatta* 8 females	Diet—for 9 months	500 ng/kg diet	Five animals died between 7 and 12 months. Epithelial tissue changes.	80
2,7-dichlorodibenzo-*p*-dioxin	Rat	Osborne-Mendel 35 males and females per group	Diet—*ad libitum* feeding	5000 and 10,000 ppm in diet	No tumors. Increased incidence of hepatotoxic lesions	155
	Mouse	B6C3F1 50 males and females per group	Diet—*ad libitum* feeding	5000 and 10,000 ppm in diet	Increased incidence of hepatoxic lesions. Increased incidence of liver tumors in male mice. No tumors in females.	155
Dibenzo-*p*-dioxin	Rat	Osborne-Mendel 35 males and females per group	Diet—*ad libitum* feeding	5000 and 10,000 ppm in diet	No tumors. Increased incidence of hepatotoxic lesions in animals on higher dose.	137
	Mouse	B6C3F1 50 males and females per group	Diet—*ad libitum* feeding	5000 and 10,000 ppm in diet	No tumors. Increased incidence of hepatotoxic lesions in animals on higher dose.	

TABLE VII. (continued)

Chemical	Animal	Species and no. used	Route of administration	Dose	Effect	Ref.
Mixture of 1,2,3,6,7,8- and 1,2,3,7,8,9-hexachlorodibenzo-*p*-dioxins (HCDD)	Rat	Osborne-Mendel 50 males and 50 females per group	By gavage 2 days/week for 104 weeks	1.25, 2.5, or 5 µg/kg/week	Not carcinogenic for male rats. Dose-related increase in incidence of hepatocellular carcinomas or neoplastic nodules in female rats.	158
	Mice	B6C3F1 50 males and 50 females per group	By gavage 2 days/week for 104 weeks	Male—1.25, 2.5, or 5 µg/kg/week Females—2.5, 5, or 10 µg/kg/week	Dose-related increase in incidence of hepatocellular carcinomas and adenomas in animals of both sexes.	158
	Mice	Swiss-Webster 30 males and 30 females per group	Dermal application in acetone suspension 3 days/week for 104 weeks, separate group pretreated with DMBA[a]	0.005 µg HCDD for 16 weeks, 0.01 µg thereafter	Not carcinogenic for animals of either sex.	159

[a] DMBA, 7,12-dimethylbenz(*a*)anthracene.

ear duct carcinoma, lymphocytic leukemia, adenocarcinoma (kidney), malignant histiocytoma, angiosarcoma (skin), and adenoma (testis). A dose-related increase in tumors was not observed, however, in animals fed TCDD concentrations 10 and 100 times this level.

A level of TCDD of 0.7 µg/kg body weight administered weekly was reported to be the threshold level for carcinogenicity in Swiss/H/Riop mice.[154] Animals dosed at 10 times this concentration had a marked reduction in life span and this probably explains the absence of a significant number of tumors in this group. At the lowest dose—0.007 µg/kg body weight—TCDD had no significant carcinogenic effect. However, in animals receiving TCDD at 0.007 µg/kg body weight in addition to the herbicide 2,4,5-trichlorophenoxyethanol (70 mg/kg body weight) there was a marked increase in tumors. In this case the authors attribute the carcinogenic response to the herbicide.[154] The level of TCDD which has a carcinogenic effect in Swiss mice is roughly comparable to the concentration which caused tumors in the Dow Chemical study.[151]

TCDD is also a carcinogen in mice when applied to the skin.[153] A thrice weekly application of 0.005 µg TCDD caused an increase in fibrosarcomas of the skin in female animals.[153] The same type of tumor was observed in male animals given 0.001 µg of the chemical; however, the increase was not statistically significant. The increase in fibrosarcomas in female animals receiving 0.005 µg TCDD was observed in animals either pretreated with 50 µg of dimethylbenzanthracene (DMBA) or untreated.

DMBA is itself a procarcinogen and requires metabolic activators to convert it into a carcinogen. Some chemicals enhance (promote) the carcinogenicity of others and are studied in the presence of chemicals which are known to have carcinogenic properties such as DMBA. DiGiovanni et al.[156] have studied the carcinogenicity of TCDD in the mouse using DMBA and report that TCDD will initiate tumors (i.e., cause them directly) but that its power to do so is weak.

TCDD as well as having carcinogenic properties is said to have anticarcinogenic properties too. Cohen et al.[157] report that pretreatment of the dioxin will inhibit the carcinogenicity of the procarcinogens benzo(a) pyrene [B(a)P] and DMBA when they are applied as skin tumor initiators in female Sencar mice.

Many chemical carcinogens will bind to DNA and it is thought that this coupling is an essential step in the initiation of a tumor. Thus TCDD causes a marked decrease in the *in vivo* covalent binding of DMBA to DNA but a significant increase in the binding of B(a)P to DNA. However, it is a metabolite of B(a)P which is thought to be the carcinogen and not benzo-(a)pyrene itself. Cohen et al.[157] claim that there was no evidence of the B(a)P metabolite 7,8-dihydroxy-9,10-epoxy-7,8,9,10-tetrahydrobenzo(a)-

pyrene being bound to the nucleic acid guanine in DNA, in TCDD-treated mice. This, they believe, may explain the tumor-inhibitory properties of TCDD which they observed.

The feeding studies in primates have not been of sufficient duration to investigate the carcinogenic properties of TCDD. Allen et al.[80] report, however, that a diet containing 500 ng TCDD/kg food is lethal for primates and that the chemical will produce hypertrophy, hyperplasia, and metaplasia of the epithelium in the bronchial tree, bile ducts, pancreatic ducts, salivary gland ducts, and the mucous membranes surrounding the eyelids.

Other polychlorinated dibenzo-p-dioxins have been tested for carcinogenicity in animals. A recent National Cancer Institute study in the United States has concluded that 2,7-dichlorodibenzo-p-dioxin is a weak carcinogen in mice. The dioxin at concentrations of 5000 and 10,000 ppm in the diet caused an increase of hepatocellular adenomas and carcinomas in male mice.[155]

A mixture of two hexachlorodibenzo-p-dioxin isomers (1,2,3,6,7,8- and 1,2,3,7,8,9-) is also carcinogenic in mice when given by gavage for 104 weeks. The isomers cause liver tumors in mice of both sexes. In rats, treated in the same manner, HCDD only causes liver cancer in female animals; the isomers were not carcinogenic for male rats under the specific test conditions.[158] Similarly when a mixture of the two isomers was painted on the skin of mice for 104 weeks the chemicals were not found to be carcinogenic.[159]

Just as the presence of chlorine atoms on the dibenzo-p-dioxin ring structure affects the toxicity of this class of compounds, so too does it affect its carcinogenic properties. The unsubstituted dibenzo-p-dioxin (DBD) is the least toxic of the 75 isomers; it is also noncarcinogenic in animal studies according to a recent report. Concentrations of 5000 and 10,000 ppm of DBD in the diet had no effect on the incidence of tumors in rats or mice.[137]

In view of TCDD being a carcinogen in animals the possibility exists that it is a potential carcinogen in humans. Several studies have investigated this possibility. Three of the investigations involved workers exposed to the herbicide 2,4,5-T. The dioxin content of the herbicide—although thought to have been low—was not measured for any of the surveys. In the first study of Finnish herbicide sprayers—who used 2,4,5-T to clear vegetation from roadsides, railway tracks, along powerlines, and in forest plantations—a higher mortality from cancer was not observed. In most instances the latency period between exposure to the herbicide and the investigation was less than 20 years.[160] Similar negative results have been obtained following a survey of Swedish railroad workers; the use of 2,4,5-T was not associated with an increased incidence of cancer.[161] The third mortality study comes to the opposite conclusion. According to Hardell and Sandstrom[162] retrospective investigation of the incidence of cancer in workers employed in Sweden's lumber industry suggests that exposure to 2,4,5-T or 2,4,5-trichlorophenol 6-27 years previously caused a sixfold increase in soft tissue sarcomas. Exposure to the herbicide or chlorinated phenols was assessed by respondents completing a questionnaire. Care was taken to avoid reference to exposure to any particular chemical, or to phenoxy herbicides in particular.[162] The difficulty of eliminating confounding factors in such a survey is well known. In view of its important conclusions the survey is being reassessed by the Swedish authorities in the event that they will be required to take some action.[163]

Others have also claimed that exposure to TCDD-contaminated 2,4,5-T-based herbicides may cause cancer. Tung[164] has said that the use of the herbicide Agent Orange by United States military forces in Vietnam may be associated with the increase in primary liver cancer in the country. Controlled epidemiological surveys to support this contention have not been conducted. The latency period between exposure to the herbicides and the noted increase in liver cancer is less than ten years. As most known carcinogens have a latency period of 20-30 years, the link between TCDD exposure and liver cancer in Vietnam is thought by many scientists to be slim.[119]

Exposure to TCDD through contact with 2,4,5-T-based herbicides is considered to be of a lower order of magnitude than that experienced by workers exposed to the dioxin in industrial accidents. Two mortality surveys of workers exposed to TCDD in industry over 25 years ago have been published. The larger of the two concerns 121 workers exposed to TCDD at Monsanto's West Virginia chemical complex in 1949.[165] The authors of the survey claim that there was no apparent excess of deaths from cancer in this cohort of workers. Deaths from cancers of the lung and lymphatic and haematopoietic system were higher than the norm, but those from the digestive and genitourinary organs were lower than expected in the cohort.[165]

The second industrial survey reports an increase in the incidence of cancers in the digestive tract. Four deaths from cancers of the gastrointestinal tract have been recorded in workers exposed to TCDD in 1953 at the Ludwigshaven site of the German chemical company BASF; only one or two cases were expected.[166] The two excess cancers were noted in workers aged 65-69. The two deaths in this age range are ten times higher than expected.[166] Clusters of various types of cancer do occur by chance and it remains to be seen whether this high incidence of intestinal cancers is a chance observation and directly attributable to TCDD exposure.

Biochemical Effects

Reference has been made to TCDD's ability to alter lipid metabolism in animals and humans. The impaired fat metabolism in TCDD-dosed animals is thought to be due to reduced transport from the liver and increased mobilization from adipose tissue.[167] A single dose of 20 μg TCDD/kg body weight causes hypercholesterolemia in rats as well as increasing the level of high-density lipoprotein (HDL).[168] The cholesterol and HDL levels rose together to reach maxima some 30 days after the TCDD dose, but the subsequent fall in the HDL level was considerably faster than for cholesterol. Diet-induced hyperlipidemias are known to reduce rather than increase HDL levels,[169] and it is thought that the increase noted in this experiment may be due to increased liver protein synthesis.[168] The changes in the HDL levels following TCDD are similar to those seen in rats of the same strain with streptozotoxin-induced diabetes.[168]

Increases in plasma cholesterol have been noted in guinea pigs fed 1 μg TCDD/kg body weight for 2 weeks.[25] Other significant changes noted in these animals were increases in plasma triglycerides, albumin, total protein, iron, and urea nitrogen. The albumin and total protein increases occurred after 7 days and were accompanied by increases in the packed cell volume, a change which the authors attribute to a general state of dehydration followed by subsequent hemoconcentration.[25] Courtney et al.[170] have shown that dehydration does occur in TCDD-dosed animals and that the primary effect of the dioxin is the cessation of water and food consumption.

Perhaps the best-known property of TCDD is its ability to induce enzyme activity in various tissues. The dioxin is a potent inducer of hepatic enzymes such as δ-aminolevulinic acid synthetase—the initial and rate limiting enzyme in the heme biosynthetic pathway[171]—the cytochrome P-450-mediated monooxygenase aryl hydrocarbon hydroxylase (AHH),[172] and 7-ethoxycoumarin o-deethylase.[173]

A single oral dose of 25 μg TCDD/kg body weight produced a marked proliferation of the smooth endoplastic reticulum in the renal proximal tubule cells, an increase which was accompanied by a considerable induction of the microsomal enzymes glucuronyl transferase and benzo(a)pyrene hydroxylase. The enzyme levels were elevated for 16 days following treatment.[174]

Enzyme induction also occurs in fetal tissue. Treatment of pregnant rats with TCDD resulted in a dose-dependent induction of AHH activity in fetal and maternal tissues. A single intraperitoneal dose of 6 μg/kg body weight produced a 24-, 22-, and fourfold increase in enzyme activity in fetal lung, kidney, and skin, respectively, while in maternal lung, kidney, and adrenals AHH activity was only increased four-, two, and twofold, respectively.[175] In an earlier experiment it was observed that a single introperitoneal injection of 2.5 μg/kg to a pregnant rat caused gross cellular damage to the placenta and the fetal and maternal livers. The cellular damage was accompanied in this case by increases in AHH activity of 100- and 14-fold in fetal and maternal livers, respectively.[176]

One effect of the enzyme induction brought about by TCDD is the alteration of hormone metabolism. Three hormones, in particular 4-androstene- 3, 17-dione, 5α-androstane-3α, 17β-diol and 4 pregnene-3,20-dione have an altered metabolism in rats injected daily for 4 days with 20 μg TCDD/kg body weight. Although metabolism was decreased in some cases, in general the metabolism of the hormones was increased several fold, the changes in female rats being higher than that in males.[177] The authors suggest that these induced changes in steroid metabolism may explain some of the signs of endocrine disturbance observed in workers exposed to TCDD such as hairfall, hirsutism, or diminished libido.[70]

All polychlorinated dibenzo-*p*-dioxins will induce AHH activity, the extent of the induction depending on the particular dioxin isomer. Poland and Kende,[171] using chick embryo liver, have shown that the potency of some 40 chlorinated dioxin isomers to induce AHH activity correlates well with the toxicity of these chemicals where known. 3-Methylcholanthrene (3-MC) is a well-known inducer of AHH activity, and in Sprague-Dawley rats 3-MC and TCDD evoke similar responses in AHH activity, the dioxin being some 30,000 times more toxic on a molar basis than 3-MC.[171]

The induction of AHH activity varies in different strains of mice. In those described as responsive, TCDD accumulation in the liver is greater than in the nonresponsive species. Furthermore the concentration of TCDD required to produce half maximal induction of AHH activity in responsive mice is about 1×10^{-9} mol/kg, an order of magnitude lower than that required to induce the same change in enzyme activity in nonresponsive animals.[171]

Increase in enzyme activity through *de novo* protein synthesis requires (1) expression (or depression) of the particular gene sequence on nuclear DNA; (2) the transcription of this DNA leading to formation of messenger RNA, and (3) the translation of the messenger to form the new protein, or enzyme. Control of the gene sequence is exercised at specific foci. The Ah locus controls the gene expressing AHH activity; it controls other genes too.[178]

TCDD, according to Poland *et al.*[178] enters the cell, binds to a specific cytosolic receptor protein, and this in turn moves into the nucleus where receptor plus TCDD are attached to DNA at the Ah locus. The binding at the Ah locus is the event which triggers the sequence of steps leading to the synthesis of AHH.

The cytosolic receptor for TCDD is highly specific for the dioxin. TCDD is bound to the protein with high affinity and the binding is both saturable and reversible.[171] Although the binding affinity of the receptor varies in different species of mouse[171] its concentration in the liver cytosol of Sprague-Dawley rats is not dependent on the sex of the animals.[179] The receptor concentration was higher, however, in newborn, 21-, and 42-day-old rats than in 7- or 56-day-old animals. Orchiectomy, ovariectomy, adrenalectomy, or hypophysectomy did not alter liver receptor cytosol concentrations significantly.[179]

Greenlee and Poland[180] have also shown that the nuclear uptake of radiolabeled TCDD is fivefold greater from the cytosol of responsive C57BL/6J mice than it is from the nonresponsive DBA/2J mouse strain. The affinity of the receptor for TCDD is ten times higher in the C57 than in the DBA strain. Uptake of TCDD by the nucleus was accompanied by a fall in the cytosol receptor concentration and both events preceded a sixfold stimulation of the activity of the microsomal enzyme 7-ethoxycoumarin O-deethylase. Hepatic uptake of ^{3}H-TCDD in C57 BL/6J mice is reduced in the presence of excess unlabeled TCDD, but unaffected by the presence of pyrene, anthracene, and phenobarbital compounds, which although enzyme inducers do not alter the enzyme profile seen with TCDD. Thus the authors argue that nuclear uptake and binding of TCDD is mediated by the cytosol receptor and that binding to the nucleus is necessary for the induction of enzyme activity.

The cytosolic receptor for TCDD is a large molecule with a molecular weight of about 136,000[181] and a sedimentation constant of between 5.0 and 6.0s,[181,182] and it is suggested[180] that the cytosolic receptor is required for the binding of TCDD to DNA. Others report the same finding and stress that the receptor must first bind TCDD before it can interact with nuclear chromatin. There is general agreement that both the cytosolic and nuclear receptors for TCDD are of a similar size and that both are sensitive to proteolytic enzymes, such as trypsin.[181,182] However, there is disagreement about whether the TCDD-cytosol receptor behaves in a similar way to known steroid receptors. With steroid receptors, translocation of the receptor to the nucleus is a temperature-dependent process. One laboratory claims that the movement of the TCDD receptor into the nucleus for binding to DNA is both temperature and time dependent.[182] Another disagrees. Binding of the TCDD receptor complex to DNA occurs without either heat activation or incubation at high ionic strength, claim Carlstedt-Duke et al.[181]

The ultimate biochemical lesion which is responsible for the toxicity of TCDD and the other polychlorinated dibenzo-p-dioxins is still not known. Neal et al.[183] exclude the following as mechanisms by which TCDD exerts its toxic effect: an adverse effect on DNA synthesis; protein synthesis; ATP production; the oxidation–reduction state of cells; the action of thyroid and glucocorticoid hormones; the general absorption of carbohydrates, amino acids, and fatty acids from the gut, their subsequent transport through cell membranes, and their metabolism. Neal et al.[183] are in agreement with Greenlee and Poland[180] that it is the cytosolic binding of TCDD to a receptor, the transferral of this complex to the nucleus, and the subsequent synthesis of critical protein(s) or enzyme(s) which could be the toxic lesion caused by TCDD. Of interest in this regard is the demonstration by Poland and Glover[184] that there is a good correlation between the binding affinity of the receptor for TCDD and the toxic response of the dioxin in animals. Atrophy of the thymus is one of the most common lesions of TCDD poisoning. A dose-response study for TCDD-induced thymic involution showed that C57BL/6J mice, which have a high affinity receptor for TCDD and are sensitive to enzyme induction, are ten times more sensitive to thymic involution than the DBA/2J mice, which have a lower affinity receptor and are less sensitive to induction of AHH activity. A hybrid B6D2F/J strain of the two species is intermediate between the two in its response to TCDD.[184]

The overall effect of TCDD is one in which the cell-mediated immune response of animals is reduced and their resistance to infection increased.[54] This effect, it appears, occurs because fewer macrophages and spleen cells can be recovered from TCDD-treated mice; the specific immune responses such as the macrophage-mediated and natural killer-cell-mediated cytotoxicity on tumor cells is not altered on a per cell basis.[185] It is the loss of cells which is responsible for the lowering of resistance to infection.

The capacity of other halogenated hydrocarbons to produce thymic atrophy corresponded to their binding affinity for the TCDD receptor.[184]

A single dose of 30 μg/kg of TCDD on day 10 of pregnancy produced an incidence of cleft palate in 54% of C57BL/6J fetuses, 13% in the B6D2F/J hybrid fetuses, and only 2% in the DBA/2J fetuses. The same dose of TCDD produced an incidence of only 0%–3% of cleft palates in five inbred strains of mice with the low-affinity receptor, while the incidence was over 5% in inbred strains with the high-affinity binding protein.[184]

TCDD is also a carcinogen. And Poland and Glover[186] have shown that the quantity of TCDD attached to nuclear DNA, i.e., nonextractable, is very small and some 4–6 orders of magnitude lower than that of most chemical carcinogens. The amount bound is 6.2 nmol TCDD per molecule of DNA, equivalent to one molecule of TCDD bound to the DNA in 35 cells. The authors argue, therefore, that covalent binding of TCDD is unlikely to be the mechanism of TCDD-induced carcinogenesis. They suggest that the carcinogenic effect of the chemical may be by acting as a promoter and thus accelerating development of a tumor only at an embryo stage, or by trophically stimulating target tissues—which contain the TCDD-cytosolic receptor—or alternatively they say that the induction of microsomal monoxygenase activity by TCDD may increase the rate of metabolism of procarcinogens to carcinogens and thus enhance their carcinogenic potential.

More recently it has been shown that TCDD is a potent promoting agent for hepatocarcinogenesis. Female Charles-Rivers rats, which were subjected to a 70% hepatectomy to stimulate cell division and given a single 10-mg/kg dose of the carcinogen diethylnitrosamine (DEN), developed tumors when given TCDD at two doses equivalent to 0.01 and 0.1 µg/kg/day. Hepatectomized animals which received only a single initiating dose of DEN and no further treatment, or which received TCDD alone, had no tumors.[187] TCDD in this experiment was clearly allowing tumors caused by DEN to develop.

However, it is known that TCDD is not just a promoter but that it is a complete carcinogen, both initiating and causing tumors to develop.[152] Poland and Glover[186] rule out a carcinogenic effect by TCDD through somatic mutation in cells. This conclusion may, however, be a little premature. While it is true that the evidence of a mutagenic effect of TCDD in the *in vitro Salmonella typhimurium* tests referred to earlier is equivocal at best, the chemical is certainly a mutagen in a test using a mammalian cell line.[134]

Thus, TCDD's mode of action as a carcinogen is still subject to dispute. Equally challenging is the need to know the biochemical lesion(s) which is responsible for the chemical's toxic action in cells. It is clear that there is still much to do.

References

1. National Academy of Sciences, *The Effects of Herbicides in South Vietnam*. Part A. Summary and Conclusions (1974).
2. Arnold, E. L., Young, A. L., and Wachnisk, A. M. Three years of field studies on the soil persistence and movement of 2,4-D, 2,4,5-T and TCDD, Abstract No. 206, Weed Society of America, 1976.
3. Kearney, P. C., Isensee, A. R., Helling, C. S., Woolson, E. A., and Plimmer, J. R. Environmental significance of chlorodioxins, *Advances in Chemistry Series* **120**, 105–111 (1973).
4. Kearney, P. C., Woolson, E. A., Isensee, A. R., and Helling, C. S. Tetrachlorodibenzodioxin in the environment: sources, fate and decontamination, *Environmental Health Perspectives* **5**, 273–277 (1973).
5. Matsumura, F., and Benezet, H. J. Studies on the bioaccumulation and microbial degradation of 2,3,7,8-tetrachlorodibenzoparadioxin, *Environmental Health Perspectives* **5**, 253–258 (1973).
6. Young, A. L., Thalken, C. E., Arnold, E. L., Cupello, J. M., and Cockerman, L. G. Fate of 2,3,7,8-tetrachlorodibenzo-*p*-dioxin (TCDD) in the environment: Summary and decontamination recommendations, USAFA-TR-76-18, Department of Chemistry and Biological Sciences, USAF Academy, Colorado 80840.
7. Seveso. Official report of the Italian Parliamentary Commission of Enquiry, 25 July 1978, p. 169.
8. Mercier, M. J., Roberfroid, M. B., Gerlache, J. de and Lans, M. 2,3,7,8-Tetrachlorodibenzoparadioxin: An overview, Document No. V/F/2499/76c; 1-50, Commission of the European Communities, Health and Safety Directorate, 1976.

9. Buu-Hoi, N. P., Saint-Ruf, G., Bigot, P., and Mangane, M. Preparation, properties and identification of dioxin (2,3,7,8-tetrachlorodibenzo-*p*-dioxin) in the pyrolysate of defoliants containing 2,4,5-T and its esters in contaminated vegetation, *Secinces Academe Science Series D* **273**, 708–711 (1971) (French).
10. Saint-Ruf, G. Formation of dioxin in the pyrolosis of sodium 2-(2,4,5-trichlorophenoxy)-propionate, *Naturwissenschaften* **59**, 648 (1972).
11. Langer, H. G., Brady, T. P., and Briggs, P. R. Formation of dibenzodioxins and other condensation products from chlorinated phenols and derivatives, *Environmental Health Perspectives* **5**, 3–7 (1973).
12. Stehl, R. H., and Lamparski, L. L. Combustion of several 2,4,5-trichlorophenoxy compounds: Formation of 2,3,7,8-tetrachlorodibenzo-*p*-dioxin, *Science* **197**, 1008–1009 (1977).
13. Ahling, B., Lindskog, A., Jansson, B., and Sundstrom, G. Formation of polychlorinated dibenzo-*p*-dioxins and dibenzofurans during combustion of a 2,4,5-T formulation, *Chemosphere* **33**, 461–468 (1977).
14. Cutler, M. R. Remarks to the national USDA/EPA Symposium on the use of herbicides in forestry. USDA-SEA Washington, D.C. 7p. Quoted by Young, A. L. *et al.* in The toxicology, environmental fate, and human risk of herbicide orange and its associated dioxin, Report OEHL TR-78-92, USAF (1978).
15. Plimmer, J. R., Klingebiel, U. I., Crosby, D. G., and Wong, A. S. Photochemistry of dibenzo-*p*-dioxins, *Advances in Chemistry Series*, by E. H. Blair, **120**, 44–54 (1973).
16. Crosby, D. G., and Wong, A. S. Environmental degradation of 2,3,7,8-tetrachlorodibenzo-*p*-dioxin (TCDD), *Science* **195**, 1337–1338 (1977).
17. Wipf, H. K., Homberger, E., Neuner, N. and Schenker, F. Field trials on photodegradation of TCDD on vegetation after spraying with vegetable oil, in *Dioxin: Toxicological and Chemical Aspects*, ed. F. Cattabeni, A. Cavallaro, and G. Galli, SP Medical and Scientific Books, New York and London (1978), pp. 201–207.
18. Nash, R. G., and Beall, M. L., Jr. Environmental distribution of 2,3,7,8-tetrachlorodibenzo-*p*-dioxin (TCDD) applied with silvex to turf in microagroecosystem, Final Report EPA-1AG-D6-0054. Agricultural Environmental Quality Institute, U.S. Department of Agriculture, Beltsville, Maryland (1978).
19. Kearney, P. C., Woolson, E. A., Isensee, A. R., and Helling, C. S. Tetrachlorodibenzodioxin in the environment: sources, fate and decontamination, *Environmental Health Perspectives* **5**, 273–277 (1973).
20. Ayres, D. C. Destruction of polychlorodibenzo-*p*-dioxins, *Nature* **290**, 323–324 (1981).
21. Piper, W. N., Rose, J. Q., and Gehring, P. J. Excretion and tissue distribution of 2,3,7,8-tetrachlorodibenzo-*p*-dioxin in the rat, *Environmental Health Perspectives* **5**, 241–244 (1973).
22. Allen, J. R., Van Miller, J. P., and Norbach, D. H. Tissue distribution, excretion and biological effects of (^{14}C) tetrachlorodibenzo-*p*-dioxin in rats, *Food and Cosmetic Toxicology* **13**, 501–505 (1976).
23. Fries, G. F., and Marrow, G. S. Retention and excretion of 2,3,7,8-tetrachlorodibenzo-*p*-dioxin by rats, *Journal of Agriculture and Food Chemistry* **23**, 265–269 (1975).
24. Rose, J. Q., Ramsey, J. C., Wentzler, T. H., Hummel, R. A. and Gehring, P. J. The fate of 2,3,7,8-tetrachlorodibenzo-*p*-dioxin following single and repeated oral doses to the rat, *Toxicology and Applied Pharmacology* **36**, 209–226 (1976).
25. Gasiewicz, T. A., and Neal, R. A. 2,3,7,8-Tetrachlorodibenzo-*p*-dioxin tissue distribution, excretion and effects on clinical chemical parameters in guinea pigs, *Toxicology and Applied Pharmacology* **51**, 329–339 (1979).
26. Poigner, H., and Schlatter, C. L. Influence of solvents and adsorbents on dermal and intestinal absorption of TCDD, *Food and Cosmetic Toxicology* (in press).

27. Young, A. L., Lehn, P. J., and Mettee, M. F. Absence of TCDD toxicity in an aquatic ecosystem, Weed Science Society America meeting, Abstract No. 107, (1976), p. 46.
28. Anonymous. Dioxin: Position Document (draft), Dioxin Working Group, U.S. Environmental Protection Agency, Washington, D.C. 17 pp. (April 1977).
29. Ross, R. T. 2,4,5-T/Dioxins: Analytical Collaborators Meeting, 15 June 1976. 25 June 1976 Memorandum of the U.S. Environmental Protection Agency, Washington, D.C. (1976), p. 3.
30. Allen, J. R., and Carstens, L. A. Light and detection microscopic observations in *Macaca mulatta* monkeys fed toxic fat, *American Journal of Veterinary Research* **28**, 1513-1526 (1967).
31. Baughman, R. W. Tetrachlorodibenzo-*p*-dioxins in the environment. High resolution mass spectrometry at the picogram level, Ph.D. thesis, Department of Chemistry, Harvard University, Cambridge, Massachusetts (1974).
32. Shadoff, L. A., Hummel, R. A., and Lamparski, L. A search for 2,3,7,8-tetrachlorodibenzo-*p*-dioxin (TCDD) in an environment exposed annually to 2,4,5-trichlorophenoxyacetic acid ester (2,4,5-T) herbicides, *Bulletin of Environmental Contamination and Toxicology* **18**, 478-485 (1977).
33. Anonymous. Dioxin traces found in soldiers exposed to defoliant, *Nature* **282**, 772 (1979).
34. Poiger, H., and Schlatter, C. L. Biological degradation of TCDD in rats. *Nature* **281**, 706-707 (1979).
35. Poiger, H., and Schlatter, C. L. Special aspects of metabolism and kinetics of TCDD, presented at workshop on "Impact of chlorinated dioxins and related compounds on the environment," Rome, 22-24 October 1980.
36. Coccia, P., Croci, T., and Manara, L. Less TCDD persists in liver 2 weeks after single dose to mice fed chow with added charcoal or cholic acid, *British Journal of Pharmacology* (in press).
37. McConnell, E. E., Moore, J. A., Haseman, J. K., and Harris, M. W. The comparative toxicity of chlorinated dibenzo-*p*-dioxins in mice and guinea pigs, *Toxicology and Applied Pharmacology* **44**, 335-356 (1978).
38. Moore, J. A., McConnell, E. E., Dalgard, D. W., and Harris, M. W. Comparative toxicity of three halogenated dibenzofurans in guinea pigs, mice, and rhesus monkeys. *Annals of the New York Academy of Sciences* **320**, 151-163 (1979).
39. Schwetz, B. A., Norris, J. M., Sparschu, G. L., Rowe, U. K., Gehring, P. O., Emerson, J. L., and Gerbig, C. G. Toxicology of chlorinated dioxins, *Advances in Chemistry Series*, ed. E. H. Blair, **120**, 55-69 (1973).
40. Harris, M. W., Moore, J. A., Vos, J. G., and Gupta, B. N. General biologic effects of TCDD in laboratory animals, *Environmental Health Perspectives* **5**, 101-109 (1973).
41. Luster, M. I., Faik, R. E., and Clark, G. Laboratory studies on the immune effects of halogenated aromatics, *Annals of the New York Academy of Sciences* **320**, 473-486 (1979).
42. McNulty, W. P. Toxicity of 2,3,7,8-tetrachlorodibenzo-*p*-dioxin for Rhesus monkeys: brief report, *Bulletin of Environmental Contamination and Toxicology* **18**, 108-109 (1977).
43. Vos, J. G., Moore, J. A., and Zinkl, J. G. Toxicity of 2,3,7,8-tetrachlorodibenzo-*p*-dioxin (TCDD) in C57BL/6 mice, *Toxicology and Applied Pharmacology* **29**, 229-241 (1974).
44. Jones, G., and Greig, J. B. Pathological changes in the livers of mice given 2,3,7,8-tetrachlorodibenzo-*p*-dioxin, *Experimentia* **31**, 1315-1317 (1975).
45. Greig, J. B., Jones, G., Butler, W. H., and Barnes, J. M. Toxic effects of 2,3,7,8-tetrachlorodibenzo-*p*-dioxin, *Food and Cosmetics Toxicology* **11**, 585-595 (1973).
46. Olson, J. R., Holscher, M. A., and Neal, R. A. Toxicity of 2,3,7,8-tetrachlorodibenzo-*p*-dioxin in the Golden Syrian Hampster, *Toxicology and Applied Pharmacology* **55**, 67-78 (1980).

47. Buu Hoi, H. P., Chanh, P. H., Sesque, G., Acum-Gelade, M. C., and Saint Ruf, G. Exzymatic functions as targets of the toxicity of dioxin 2,3,7,8-tetrachlorodibenzo-*p*-dioxin, *Natuurwissenschaften* **59**, 173-174 (1972).
48. Cunningham, M. H., and Williams, D. F. Effect of tetrachlorodibenzo-*p*-dioxin on growth rate and the synthesis of lipids and proteins in rats, *Bulletin of Environmental Contamination and Toxicology* **7**, 45-51 (1972).
49. Greig, J. B., and de Matteis, F. Effects of 2,3,7,8-tetrachlorodibenzo-*p*-dioxin on drug metabolism and hepatic microsomes of rats and mice. *Environmental Health Perspectives* **5**, 211-219 (1973).
50. Kociba, R. J., Keeler, P. A., Park, C. N., and Gehring, P. J. 2,3,7,8-tetrachlorodibenzo-*p*-dioxin (TCDD): results of a 13-week oral toxicity study in rats, *Toxicology and Applied Pharmacology* **35**, 553-574 (1976).
51. Vos, J. G., Kreeftenberg, J. G., Engel, H. W. B., Minderhoud, A., and Van Noorle Jansen, L. M. Studies on 2,3,7,8-tetrachlorodibenzo-*p*-dioxin-induced immune suppression and decreased resistance to infection; endotoxin hypersensitivity, serum zinc concentrations and effect on thymus in treatment, *Toxicology* **9**, 75-86 (1978).
52. Vos, J. G., and Moore, J. A. Suppression of cellular immunity in rats and mice by maternal treatment with 2,3,7,8-tetrachlorodibenzo-*p*-dioxin, *Internal Archives of Allergy and Applied Immunology* **47**, 777-794 (1974).
53. Greig, J. B. The toxicology of 2,3,7,8-tetrachlorodibenzo-*p*-dioxin and its structural analogues, *Annals of Occupational Hygiene* **22**, 421-427 (1979).
54. Thigpen, J. E., Faith, R. E., McConnell, E. E., and Moore, J. A. Increased susceptibility to bacterial infection as a sequela of exposure to 2,3,7,8-tetrachlorodibenzo-*p*-dioxin, *Infection and Immunology* **12**, 1319-1324 (1975).
55. Vos, J. G., 2,3,7,8-tetrachlorodibenzo-para-dioxin: effects and mechanisms, in *Chlorinated Phenoxy Acids and their Dioxins* ed. C. Ramel, *Ecological Bulletin (Stockholm)* **27**, 165-176 (1978).
56. Faith, R. E., and Luster, M. I. Investigations on the effects of 2,3,7,8-tetrachlorodibenzo-*p*-dioxin (TCDD) on parameters of various immune functions, *Annals of the New York Academy of Sciences* **320**, 564-571 (1979).
57. Milford Ward, A. Unpublished report, Investigation of the immune capability of workers previously exposed to 2,3,7,8-tetrachlorodibenzo-para-dioxin (TCDD), Supraregional Specific Protein Reference Unit and Department of Immunology, Hallamshire Hospital, Sheffield, U.K. (1978).
58. Kimmig, J., and Schultz, K. H. Berufliche Akne (sog Chlorakne) durch chlorierte aromatische zyklische Ather, *Dermatologica* **115**, 540-546 (1957).
59. Bauer, H., Schultz, K. H., and Spiegelberg, U. Berufliche Vergiftungen bei der Herstellung von Chlorophenol-Verbindungen, *Archiv fuer Gewerbepathology un Gewerbehygin* **18**, 538-555 (1961).
60. Gupta, B. N., Vos, J. G., Moore, J. A., Zinkl, J. G., and Bullock, B. C. Pathologic effects of 2,3,7,8-tetrachlorodibenzo-*p*-dioxin in laboratory animals, *Environmental Health Perspectives* **5**, 125-140 (1973).
61. Jones, G., and Butler, W. H. A morphological study of the liver lesion induced by 2,3,7,8-tetrachlorodibenzo-*p*-dioxin (dioxin) in rats, *Journal of Pathology* **112**, 93-97 (1974).
62. Truhart, R., Chanh, P. H., Van Haverbeck, G., Azum-Gelade, M. C., Saint Ruf, G., and Lareng, L. Toxicite a court terme de la tetrachloro-2,3,7,8-dibenzo-*p*-dioxin chez le rat: etude structurale, ultrastructureate et enzymologique du foie, *Comptes Rendus des Seances de la Societe de Biologie et des ses Filiales (Paris)* **279**, 1565-1569 (1974).
63. Jones, G. A histochemical study of the liver lesion induced by 2,3,7,8-tetrachlorodibenzo-*p*-dioxin (dioxin) in rats, *Journal of Pathology* **116**, 101-105 (1975).
64. Greig, J. B., and Osborne, G. Changes in rat hepatic cell membranes during 2,3,7,8-tetrachlorodibenzo-*p*-dioxin intoxication, in *Dioxin: Toxicological and Chemical Aspects*,

ed. F. Cattabeni, A. Cavallaro, and G. Galli, Spectrum Publications, New York (1978), pp. 105-111.
65. Goldstein, J. A., Hickman, P., Bergman, H., and Vos, J. G. Hepatic porphyria induced by 2,3,7,8-tetrachlorodibenzo-p-dioxin in the mouse, Research Communications in Chemistry, Pathology and Pharmacology 6, 919-928 (1973).
66. Sinclair, P. R., and Granick, S. Uroporphyrin formation induced by chlorinated hydrocarbons (lindane, polychlorinated biphenyls, tetrachlorodibenzo-p-dioxin): requirements for endogenous iron, protein synthesis and drug metabolizing activity, Biochemical, Biophical Research Communications 61, 124-133 (1974).
67. Goldstein, J. A., Hickman, P., and Bergman, H. Induction and hepatic porphyria and drug metabolizing enzymes by 2,3,7,8-tetrachlorodibenzo-p-dioxin, Federation Proceedings of the American Society Experimental Biology 35, 708 (1976).
68. Poland, A., and Glover, E. Chlorinated dibenzo-p-dioxins: potent inducers of delta-aminolevulinic acid synthetase and aryl hydrocarbon hydroxylase II. A study of the structure-activity relationship, Molecular Pharmacology 9, 736-747 (1973).
69. McConnell, E. E., and Moore, J. A. The toxicopathology of TCDD, in Dioxin: Toxicological and Chemical Aspects, ed. F. Cattabeni, A. Cavallaro, A., and G. Galli, Spectrum Publications, New York (1978), pp. 137-141.
70. Oliver R. M. Toxic effects of 2,3,7,8-tetrachlorodibenzo-1,4-dioxin in laboratory workers, British Journal of Industrial Medicine 32, 49-53 (1975).
71. Bleiberg, J., Wallen, M., Brodkin, R., and Appelbaum, I. L. Industrially acquired porphyria, Archives Dermatology 89, 793-797 (1964).
72. Jirasek, L., Kalensaky, J., Kubec, K., Pazderova, J., and Lukas, E. Acne chlorina, porphyria cutanea tarda and other manifestations of general intoxication during the manufacture of herbicides, Ceskoslovenska Dermatologie 49, 145-147 (1974).
73. Zinkl, J. G., Vos, J. G., Moore, J. A., and Gupta, B. N. Haematologic and clinical chemistry effects of 2,3,7,8-tetrachlorodibenzo-p-dioxin in laboratory animals, Environmental Health Perspectives 5, 111-118 (1973).
74. Poland, A., Smith, D., Mether, G., and Possick, P. A health survey of workers in a 2,4-D and 2,4,5-T plant, Archives of Environmental Health 22, 316-327 (1971).
75. May, G. Chloracne from the accidental production of tetrachlorodibenzo-p-dioxin, British Journal of Industrial Medicine 30, 276-283 (1973).
76. Sanger, V. L., Scott, L., Hamdry, A., Gale, C., and Pounden, N. D. Alimentary toxemia in chickens, Journal of American Veterinary Medical Association 133, 172-176 (1958).
77. Cantrell, J. S., Webb, N. C., and Mabis, A. J. Search for the chick edema factor, Chemical Engineering News 45, 10 (1967).
78. Firestone, D. Etiology of chick edema disease, Environmental Health Perspectives 5, 59-66 (1973).
79. Khera, K. A., and Ruddick, J. A. Polychlorodibenzo-p-dioxins: perinatal effects and the dominant lethal test in Wistar rats, in Chlorodioxins—Origins and Fate, Advances in Chemistry Series, ed. E. H. Blair, 120, 70-84 (1973).
80. Allen, J. R., Barsotti, D. A., Van Miller, J. P., Abrahamson, L. J., and Lalich, J. J. Morphological changes in monkeys consuming a diet containing low levels of 2,3,7,8-tetrachlorodibenzo-p-dioxin, Food and Cosmetic Toxicology 15, 401-410 (1977).
81. Carter, C. D., Kimbrough, R. D., Liddle, J. A., Cline, R. E., Zack, M. M. Barthel, W. R., Koehler, R. E., and Philips, P. E. Tetrachlorodibenzodioxin: an accidental poisoning episode in horse arena, Science 188, 738-740 (1975).
82. Miller, R. A., Norris, L. A., and Hawkes, C. L. Toxicity of 2,3,7,8-tetrachlorodibenzo-p-dioxin in aquatic organisms, Environmental Health Perspectives 5, 177-186 (1973).
83. Norris, L. A., and Miller, R. A. The toxicity of 2,3,7,8-tetrachlorodibenzo-p-dioxin (TCDD) in Guppies (Poecilia reticulatus, Peters), Bulletin of Environmental Contamination and Toxicology 12, 76-80 (1974).

84. Schultz, K. H. Zur Klinic und Atiologie und Atiologie der Chlorakne, *Arbeitsmedizin Socialmedizin Arbeitshygiene* **2**, 25-29 (1968).
85. Telegina, K. A., and Bikbulatova, L. J. Affection of the follicular apparatus of the skin of workers employed in the production of the butyl ester of 2,4,5-trichlorophenoxyacetic acid (in Russian), *Vestnik Dermatologis Venerdogii (Moskva)* **44**, 35-39 (1970).
86. Hay, A. Toxic cloud over Seveso, *Nature (London)* **262**, 636-638 (1976).
87. Hay, A. Seveso: the aftermath, *Nature (London)* **263**, 538-540 (1976).
88. Dugois, P., and Colombe, L. Remarques sur l'acne chlorique (a propos d'une eclosion de cas provoques par la preparation du 2,4,5-trichlorophenol), *Le Journal de Medicin de Lyon* **38**, 899-903 (1957).
89. Reggiani, G. Health risks at Seveso, paper presented at NATO Ecotoxicology Workshop, University of Surrey, Guildford, U.K., 1-5 August 1977.
90. Reggiani, G. Acute human exposure to TCDD in Seveso, Italy, *Journal of Toxicology and Environmental Health* **6**, 27-43 (1980).
91. Reggiani, G. Estimation of the TCDD toxic potential in the light of the Seveso accident, *Archives of Toxicology*, Supplement 2, 291-302 (1979).
92. Homberger, E., Reggiani, G., Sambeth, J., and Wipf, H. K. The Seveso Accident: its nature, extent and consequences, *Annals of Occupational Hygiene* **22**, 327-370 (1979).
93. Reggiani, G. Localized contamination with TCDD—Seveso, Missouri and other areas, in *Halogenated Biphenyls, Terphenyls, Napthalenes, Dibenzodioxins and Related Products*, ed. R. Kimbrough, Elsevier; North Holland Biomedical Press (1980), pp. 301-317.
94. Martin, J. V. Report on a biochemical study carried out on workers at Coalite and Chemical Products Ltd., Derbyshire, England (unpublished).
95. Walker, A. L., and Martin, J. V. Lipid profiles in dioxin-exposed workers, *Lancet* **I**, 446 (1979).
96. Neubert, D., and Dillmann, I. Embryotoxic effects in mice treated with 2,4,5-trichlorophenoxyacetic acid and 2,3,7,8-tetrachlorodibenzo-*p*-dioxin, *Naunym-Schmiedebergs Archives of Pharmacology (Berlin)* **272**(3), 243-264 (1972).
97. Neubert, D., Zens, P., Rothenwaller, A., and Merker, H. J. A survey of the embryotoxic effects of TCDD in mammalian species, *Environmental Health Perspectives* **5**, 67-79 (1973).
98. Courtney, K. D., and Moore, J. A. Teratology studies with 2,4,5-trichlorophenoxyacetic acid and 2,3,7,8-tetrachlorodibenzo-*p*-dioxin, *Toxicology and Applied Pharmacology* **20**, 396-403 (1971).
99. Courtney, K. D. Mouse teratology studies with chlorodibenzo-*p*-dioxin, *Bulletin of Environmental Contamination and Toxicology* **16**, 674-681 (1976).
100. Smith, F. A., Schwetz, B. A., and Nitsche, D. K. Teratogenicity of 2,3,7,8-tetrachlorodibenzo-*p*-dioxin in CF-1 mice, *Toxicology and Applied Pharmacology* **38**, 517-523 (1976).
101. Moore, J. A., Gupta, B. N., Zinkl, J. G., and Vos, J. G. Postnatal effects of maternal exposure to 2,3,7,8-tetrachlorodibenzo-*p*-dioxin (TCDD), *Environmental Health Perspectives* **5**, 81-85 (1973).
102. Sparschu, G. L., Dunn, F. L., and Rowe, V. K. Study of the teratogenicity of 2,3,7,8-tetrachlorodibenzo-*p*-dioxin in the rat, *Food and Cosmetic Toxicology* **9**, 405-412 (1971).
103. Roll, R. Studies of the teratogenic effect of 2,4,5-T in mice, *Food and Cosmetic Toxicology* **9**, 671-676 (1971).
104. Courtney, K. D., Gaylor, D. W., Hogan, M. D., Falk, H. L., Bates, R. R. and Mitchell, I. Teratogenic evaluation of 2,4,5-T, *Science* **168**, 864-866 (1970).
105. Nau, H., and Bass, R. Transfer of 2,3,7,8-tetrachlorodibenzo-*p*-dioxin (TCDD) to the mouse embryo and fetus, presented at workshop on "Impact of Chlorinated Dioxins and Related Compounds on the Environment," Rome, 22-24 October 1980.
106. Sparschu, G. L., Dunn, F. L., Lisowe, R. W., and Rowe, V. K. Study of the effects of

high levels of 2,4,5-trichlorophenoxyacetic acid on fetal development in the rat, *Food and Cosmetic Toxicology* **9**, 527–530 (1971).
107. Sokolik, I. Yu. Effect of 2,4,5-trichlorophenoxyacetic acid and its butyl ester on embryogenesis of rats, *Byulletin eksperimental'noi biologii i meditsiny* **76**, 90–92 (1973).
108. Smith, F. A., Murray, F. J., John, J. A., Nicschke, K. D., Kociba, R. J., and Schwetz, B. A. Three-genration reproduction study of rats ingesting 2,4,5-trichlorophenoxyacetic acid in the diet, *Food and Cosmetic Toxicology* **19**, 41–45 (1981).
109. Collins, T. F. X., and Williams, C. H. Teratogenic studies with 2,4,5-T and 2,4-D in the hamster, *Bulletin of Environmental Contamination and Toxicology* **6**, 559–567 (1971).
110. Emerson, J. L., Thompson, D. J., Strebing, R. J., Gerbig, C. J., and Robinson, V. B. Teratogenic studies on 2,4,5-trichlorophenoxyacetic acid in the rat and rabbit, *Food and Cosmetic Toxicology* **9**, 395–404 (1971).
111. *Advisory Committee on 2,4,5-T*. Report of the Advisory Committee on 2,4,5-T to the administrator of the Environmental Protection Agency 76 pp. (1971).
112. Binns, W., and Balls, L. Non-teratogenic effects of 2,4,5-trichlorophenoxyacetic acid and 2,4,5-T propylene glycol butyl ester herbicides in sheep, *Teratology* **4**, 245 (1971).
113. 2,4,5-T and Human Birth Defects, New Zealand Department of Health (June 1977).
114. Consultative Council on Congenital Abnormalities in the Yarram district, Victoria, Australia, Department of Primary Industry, Canberra, Report (1978).
115. Anonymous press statement. Australian National Health and Medical Research Council (16 June 1978).
116. Hickey, G. Anthropologist's report in *The Effects of Herbicides in South Vietnam*, Part A, Summary and Conclusions, National Academy of Sciences, Washington, D.C. (1974), pp. VII58–VII66.
117. Tung, T. T., Anh, T. K., Tuyen, B. Q., Tra, D. X., and Huyan, N. X. Clinical effects on the civilian population as a result of the massive and continuous use of defoliants, Preliminary survey, *Vietnamese Studies* **29**, 53–83 (1971).
118. Meselson, M. Preliminary report of Herbicide Assessment Commission of the American Association for the Advancement of Science, U.S. Congressional Record, 92 Congress, 2nd Session 118 (32): S 3226-3227 (March 3 1972).
119. Hay, A. Vietnam's dioxin problem, *Nature (London)* **271**, 597–598 (1978).
120. Alsea II, Report of assessment of a field investigation of six year spontaneous abortion rates in three Oregon areas in relation to forest 2,4,5-T spray practices, U.S. Environmental Protection Agency (28 February 1979).
121. Cookson, C. Emergency ban on 2,4,5-T herbicide in U.S., *Nature (London)* **278**, 108–109 (1979).
122. Hay, A. Critics challenge data that led to 2,4,5-T ban, *Nature (London)* **279**, 3 (1979).
123. Anonymous. Dow study shows no correlation between dioxin exposure and pregnancy results, *Dow Today*, Dow Communications, Dow Chemical Company (24 December 1980).
124. Bridges, B. Short term screening tests for carcinogens, *Nature (London)* **261**, 195–200 (1976).
125. Ames, B. N., McCann, J., and Yamasaki, E. Methods for detecting carcinogens and mutagens with the Salmonella/Mammalian—Microsome Mutagenicity test, *Mutation Research* **31**, 347–364 (1975).
126. Purchase, I. F. H., Longstaff, E., Ashby, J., Styles, J. A., Anderson, D., Lefevre, P. A., and Westwood, F. R. An evaluation of 6 short term tests for detecting organic chemical carcinogens, *British Journal of Cancer* **37**, 873–959 (1978).
127. Evaluation of short term tests for carcinogens: Report of the International Collaborative Program, ed. F. J. de Serres and J. Ashby, *Progress in Mutation Research, Vol 1*, Elsevier/North-Holland, Amsterdam (1981).
128. Ames, B. N., Lee, F. D., and Durston, W. E. An improved bacterial test system for the

detection and classification of mutagens and carcinogens, *Proceedings of the National Academy of Sciences (USA)* **70**, 782-786 (1973).
129. Hussain, S., Ehrenberg, L., Lofroth, G., and Gejvall, T. Mutagenic effects of TCDD on bacterial systems, *Ambio* **1**, 32-33 (1972).
130. Seiler, J. P. A survey of the mutagenicity of various pesticides, *Experientia* **29**, 622-623 (1973).
131. Wassam, J. S., Huff, J. E., and Loprieno, N. A review of the genetic toxicology of chlorinated dibenzo-*p*-dioxin, *Mutation Research* **47**, 141-160 (1977/8).
132. McCann, J. Reported by Wassam *et al.*, Reference 131.
133. Nebert, D. W., Thorgeirsson, S. S., and Felton, J. S. Genetic differences in mutagenesis, carcinogenesis and drug toxicity, in *In Vitro Metabolic Activation in Mutagenesis Testing*, ed. F. J. de Serres, J. R. Fouts, J. R. Bend, and R. M. Philpot, Elsevier./North-Holland, Amsterdam (1976), pp. 105-124.
134. Ashby, J., Styles, J. A., Elliott, B., and Hay, A. W. M. (unpublished).
135. Styles, J. A. A method for detecting carcinogenic organic chemicals using mammalian cells in culture, *British Journal of Cancer* **36**, 558-563 (1977).
136. Commoner, B. Reliability of bacterial mutagenesis techniques to distinguish carcinogenic and non-carcinogenic chemicals, Environmental Protection Agency Report No. EPA-600/1-76-022 (1976), pp. 1-103.
137. Anonymous. Bioassay of dibenzo-*p*-dioxin for possible carcinogenicity, National Cancer Institute Technical Report Series No. 122 (1979).
138. Ashby, J., Elliott, B., and Styles, J. A. Norharman and Ellipticine: A comparison of their abilities to interact with DNA *in vitro*, *Cancer Letters* **9**, 21-33 (1980).
139. Jackson, W. T. Regulation of mitosis III. Cytological effects of 2,4,5-trichlorophenoxyacetic acid and of dioxin contaminants in 2,4,5-T formulations, *Journal of Cell Science* **10**, 15-25 (1972).
140. Kimbrough, R. D., Carter, C. D., Liddle, J. A., Cline, R. E., and Philips, P. E. Epidemiology and pathology of a tetrachlorodibenzodioxin poisoning episode, *Archives of Environmental Health* **32**, 77-86 (1977).
141. Green, S., Moreland, F., and Sheu, C. Cytogenetic effects of 2,3,7,8-tetrachlorodibenzo-*p*-dioxin on rat bone marrow cells, *Food and Drug Administration Bylines*, No. 6, 292-294 (1977).
142. Killian, D. J. Industrial Medicine, Dow Chemical, Texas Division, 1973, (unpublished).
143. Hay, A. W. M. Tetrachlorodibenzo-*p*-dioxin release at Seveso, *Disasters* **1**, 289-308 (1977).
144. Blank, C. E. Bolsover Project 1977-8. Genetic Damage, Centre for Human Genetics Sheffield, U.K. (unpublished).
145. Van Miller, J. P., and Allen, J. R. Chronic toxicity of 2,3,7,8-tetrachlorodibenzo-*p*-dioxin in rats. *Federation of the American Society of Experimental Biology* **36**, 396 (1977).
146. Seiler, J. P. Inhibition of testicular DNA synthesis by chemical mutagens and carcinogens. Preliminary results in the validation of a novel short term test, *Mutation Research* **46**, 305-310 (1977).
147. Murray, F. J., Smith, F. A., Nitschke, Humiston, C. G., Kociba, R. J., and Schwetz, B. A. Three generatoin reproduction study of rats ingesting 2,3,7,8-tetrachlorodibenzo-*p*-dioxin, *Toxicology and Applied Pharmacology* **41**, 200-201 (1977).
148. Allen, J. R., Barsotti, D. A., Lambrecht, L. K., and Van Miller, J. P. Reproductive effects of halogenated aromatic hydrocarbons on nonhuman primates, *Annals of the New York Academy of Sciences* **320**, 419-425 (1979).
149. Knutson, J., and Poland, A. 2,3,7,8-tetrachlorodibenzo-*p*-dioxin: Failure to demonstrate toxicity in 23 cultured cell types, *Toxicology and Applied Pharmacology* **54**, 377-383 (1980).
150. Allen, J. R., Lalich, J. J., and Van Miller, J. P. Increased incidence of neoplasma in rats

exposed to low levels of 2,3,7,8-tetrachlorodibenzo-*p*-dioxin, *Chemosphere* **9**, 537–544 (1977).
151. Kociba, R. J., Kayes, D. G., Beyer, J. E., Carreon, R. M., Wade, C. E., Dittenber, D. A., Kalnins, R. P., Frauson, L. E., Park, C. N., Barnard, S. D., Hummel, R. A., and Humiston, C. G. Results of a two-year chronic toxicity and oncogenicity study of 2,3,7,8-tetrachlorodibenzo-*p*-dioxin in rats, *Toxicology and Applied Pharmacology* **46**, 279–303 (1978).
152. Anonymous. Bioassay of 2,3,7,8-tetrachlorodibenzo-*p*-dioxin for possible carcinogenicity (gavage study), National Cancer Institute DHHS Publication No. (NIH) 80-1757 (1980).
153. Anonymous. Bioassay of 2,3,7,8-tetrachlorodibenzo-*p*-dioxin for possible carcinogenicity (dermal study), National Cancer Institute DHHS Publication No. (NIH) 80-1757 (1980).
154. Toth, K., Somfai-Rell, S., Sugar, J., and Bence, J. Carcinogenicity testing of herbicide 2,4,5-trichlorophenoxyethanol containing dioxin and of pure dioxin in Swiss mice, *Nature (London)* **278**, 548–549 (1979).
155. Anonymous. Bioassay of 2,7-dichlorodibenzo-*p*-dioxin for possible carcinogenicity, National Cancer Institute Technical Report Series No. 123 (1979).
156. DiGiovanni, J., Viaje, A., Berry, D. L., Slaga, T. J., and Juchau, M. R. Tumor initiating ability of 2,3,7,8-tetrachlorodibenzo-*p*-dioxin (TCDD) and Arochlor 1254 in the two stage system of mouse skin carcinogenesis, *Bulletin of Environmental Contamination and Toxicology* **18**, 552–557 (1977).
157. Cohen, G. M., Bracken, W. M., Iyer, R. P., Berry, D. L., Selkirk, J. K., and Slaga, T. J. Anticarcinogenic effects of 2,3,7,8-tetrachlorodibenzo-*p*-dioxin on Benzo(*a*)pyrene and 7,12-dimethylbenz(*a*)anthracene tumor initiation and its relationship to DNA binding, *Cancer Research* **39**, 4027–4033 (1979).
158. Anonymous. Bioassay of a mixture of 1,2,3,6,7,8- and 1,2,3,7,8,9-hexachlorodibenzo-*p*-dioxin for possible carcinogenicity (gavage study), Natoinal Cancer Institute, Technical Report Series No. 198 (1980).
159. Anonymous. Bioassay of a mixture of 1,2,3,6,7,8- and 1,2,3,7,8,9-hexachlorodibenzo-*p*-dioxin for possible carcinogenicity (dermal study), National Cancer Institute, Technical Report Series No. 202 (1980).
160. Riihimaki, V., Asp, S., Sepalainen, A. M., and Hernberg, S. Symptomatology, morbidity and mortality experience of chlorinated phenoxyacid herbicide (2,4-D; 2,4,5-T) sprayers in Finland. A clinical and epidemiological study, presented at the IARC working group meeting on "Long Term Hazards of Chlorinated Dibenzodioxins and Chlorinated Dibenzofurans." International Agency for Research on Cancer, Lyon, France (10–11 January 1978).
161. Axelson, O., and Sundell, L. Herbicide exposure, mortality and tumor incidence. An epidemiological investigation on Swedish railroad workers, *Work and Environmental Health* **11**, 21–28 (1974).
162. Hardell, L., and Sandstrom, A. Case-control study: soft tissue sarcomas and exposure to phenoxyacetic acids or chlorophenols, *British Journal of Cancer* **39**, 711–717 (1979).
163. Anonymous. Dioxin and 2,4,5-T: What are the risks? *Nature (London)* **284**, 111 (1980).
164. Tung, T. T. Pathologie Humaine et Animale de la Dioxine, *La Revue de Medicin* **14**, 653–657 (1977).
165. Zack, J. A., and Suskind, R. R. The mortality experience of workers exposed to tetrachlorodibenzodioxin in a trichlorophenol process accident, *Journal of Occupational Medicine* **22**, 11–14 (1980).
166. Thiess, A., and Frentzel-Beyme, R. Mortality study of persons exposed to dioxin after an accident which occurred in the BASF on 13 November 1953, paper presented at 5th International Medichem Congress, San Francisco (5–9 September 1977).

167. Cunningham, H. M., and Williams, D. T. Effect of tetrachlorodibenzo-p-dioxin on growth rate and synthesis of lipids and proteins in rats, *Bulletin of Environmental Contamination and Toxicology* **7**, 45–51 (1972).
168. Poli, A., Franceschini, G., Puglisi, L., and Sirtoni, C. R. Increased total an high density lipoprotein cholesterol with apoprotein changes resembling streptozotoxin diabetes in tetrachlorodibenzodioxin (TCDD) treated rats, *Biochemical Pharmacology* **29**, 835–838 (1980).
169. Narayan, K. A. Lowered serum concentration of high density lipoproteins in cholesterol-fed rats, *Atherosclerosis* **13**, 205–215 (1971).
170. Courtney, K. D., Putnam, J. P., and Andrews, J. E. Metabolic studies with TCDD (Dioxin) treated rats, *Archives of Environmental Contamination and Toxicology* **7**, 385–396 (1978).
171. Poland, A., and Kende, A. 2,3,7,8-tetrachlorodibenzo-p-dioxin: environmental contaminant and molecular probe, *Federation Proceedings* **35**, 2404–2411 (1976).
172. Poland, A., and Glover, E. Comparison of 2,3,7,8-tetrachlorodibenzo-p-dioxin, a potent inducer of aryl hydrocarbon hydroxylase, with B-methyl cholanthrene, *Molecular Pharmacology* **10**, 349–359 (1974).
173. Greenlee, W. F., and Poland, A. An improved assay of 7-ethoxycoumarin 0-deethylase activity: induction of hepatic enzyme activity in C57BL/6J and DBA/2J mice by phenobarbital, 3-methylcholanthrene and 2,3,7,8-tetrachlorodibenzo-p-dioxin, *Journal of Pharmacology and Experimental Therapeutics* **205**, 596–605 (1978).
174. Fowler, B. A., Hook, G. E. R., and Lucier, G. W. Tetrachlorodibenzo-p-dioxin induction of renal microsomal enzymes systems: ultrastructural effects on pars recta (S_3) proximal tubule cells of the rat kidney, *Journal of Pharmacology and Experimental Therapeutics* **203**, 712–721 (1977).
175. Berry, D. L., Slaga, T. J., Wilson, N. M., Zachariah, P. K., Namking, M. J., Bracken, W. M., and Juchau, M. R. Transplancental induction of mixed function oxygenases in extra-hepatic tissues by 2,3,7,8-tetrachlorodibenzo-p-dioxin, *Biochemical Pharmacology* **26**, 1383–1388 (1977).
176. Berry, D. L., Zachariah, P. K., Namkung, M. J., and Juchau, M. R. Transplacental induction of carcinogen hydroxylating systems with 2,3,7,8-tetrachlorodibenzo-p-dioxin, *Toxicology and Applied Pharmacology* **36**, 569–584 (1976).
177. Gustafsson, J. A., and Sundberg, M. I. Changes in steroid hormone metabolism in rat liver microsomes following administration of 2,3,7,8-tetrachlorodibenzo-p-dioxin (TCDD), *Biochemical Pharmacology* **28**, 497–499 (1979).
178. Poland, A., Greenlee, W. F., and Kende, A. S. Studies on the mechanism of action of the chlorinated dibenzo-p-dioxins and related compounds, *Annals of the New York Academy of Sciences* **320**, 214–230 (1979).
179. Carlstedt-Duke, J. M. B., Elfstrom, G., Hogberg, B., and Gustafsson, J. A. Ontogeny of the rat hepatic receptor for 2,3,7,8-tetrachlorodibenzo-p-dioxin and its endocrine independence, *Cancer Research* **39**, 4653–4656 (1979).
180. Greenlee, W. F., and Poland, A. Nuclear uptake of 2,3,7,8-tetrachlorodibenzo-p-dioxin in C57Bl/6J and DBA/2J mice, *Journal of Biological Chemistry* **254**, 9814–9821 (1979).
181. Carlstedt-Duke, J. M. B., Harnemo, U. B., Hogberg, B., and Gustafsson, J. A. Interaction of the hepatic receptor protein for 2,3,7,8-tetrachlorodibenzo-p-dioxin with DNA, *Biochimica et Biophysica Acts* **672**, 131–141 (1981).
182. Okey, A. B., Bondy, G. P., Mason, M. E., Nebert, D. W., Forster-Gibson, C. J., Muncan, J., and Dufresne, M. J. Temperature-dependent cytosol-to-nucleus translocation of the Ah receptor for 2,3,7,8-tetrachlorodibenzo-p-dioxin in continuous cell culture lines, *Journal of Biological Chemistry* **255**, 11415–11422 (1980).
183. Neal, R. A., Beatty, P. W., and Gasiewicz, T. A. Studies of the mechanisms of toxicity of

2,3,7,8-tetrachlorodibenzo-*p*-dioxin (TCDD), *Annals of the New York Academy of Sciences* **320**, 204–213 (1979).
184. Poland, A., and Glover E. 2,3,7,8-tetrachlorodibenzo-*p*-dioxin: Segregation of toxicity with the Ah locus, *Molecular Pharmacology* **17**, 86–94 (1980).
185. Mantovani, A., Vecchi, A., Luini, W., Sironi, M., Cadiani, G. P., Spreafico, F., and Garattini, S. Effect of 2,3,7,8-tetrachlorodibenzo-*p*-dioxin on macrophage and natural killer cell-mediated cytotoxicity in mice, *Biomedicine* **32**, 200–204 (1980).
186. Poland, A., and Glover, E. An estimate of the maximum *in vivo* covalent binding of 2,3,7,8-tetrachlorodibenzo-*p*-dioxin to rat liver protein, ribosomal RNA, and DNA, *Cancer Research* **39**, 3341–3344 (1979).
187. Pitot, H. C., Goldsworthy, T., Campbell, H. A., and Poland, A. Quantitative evaluation of the promotion by 2,3,7,8-tetrachlorodibenzo-*p*-dioxin of hepatocarcinogenesis from diethylnitrosamine, *Cancer Research* **40**, 3616–3620 (1980).

3

Hexachlorophene

Hexachlorophene, the other major product derived from trichlorophenol (see Chapter 1), is a general poison effective in the control of bacteria classified as gram-positive. In the cosmetics industry, hexachlorophene is used as a preservative. For medical purposes hexachlorophene is used in the control of staphylococcal organisms. The bacteriacide has four main uses: treatment of acne and impetigo, cleansing of intact skin around burns and wounds, presurgical washing and cleansing of newborn infants, particularly the umbilical cord.

Its use, however, has been much reduced and there are claims that it could be dispensed with altogether. An alternative to hexachlorophene is chlorhexidene. A widely available bacteriacide, chlorhexidene is also used surgically in the United Kingdom as a skin cleanser, wound steriliser, and for presurgical washing. On the question of disinfecting hands, however, doubts have been expressed about the value of either chlorhexidene or hexachlorophene compared with plain soap and water. In a comparative study in which nine methods of hand disinfecting were compared, cultures of bacteria inoculated on the skin were almost completely removed (99.6% or more) either by rinsing with antiseptics or by washing with soap and water. There was little advantage in the use of antiseptics and neither rinsing with hypochlorite solution nor washing with hexachlorophene detergent cream had much advantage over soap and water. A chlorhexidene solution was only marginally more effective than soap and water.[1]

As can be seen from Table VIII, hexachlorophene is a toxic compound in its own right, as indeed it would have to be if its prime purpose is to kill bacteria. However, based on the relatively crude LD_{50} test, hexachlorophene is ten times more toxic than chlorhexidene. A newer introduction to the market, chlorhexidene, though it is considered potentially neurotoxic, has few toxic effects listed against it.[2] These include skin sensitivity, irritation of the conjunctiva of the eye and other sensitive tissues with strong solutions, and a warning that the bacteriacide should not be used on the "brain, meninges or perforated ear drum."

TABLE VIII. Bacteriacides

Product name	Common name	Chemical name	Method of production	Structure	Toxicity	Properties
Hexachlorophane (Givaudan) sole world producers. U.S. Patent Nos. 2,435,593 (1948) 2,812,365 (1957)	Hexachlorophene	2,2-dihydroxy-3,3,5,5,6,6-hexachlorodiphenylmethane	2,4,5-Trichlorophenol → hexachlorophene	(a)	LD_{50}—acute oral Rats—161 mg/kg Mice—168 mg/kg Dogs—40–50 mg/kg	Bactericide effective against gram-positive bacteria. Disinfectant used for skin washes, cleansing intact skin around burns and wounds, presurgical washes, treatment of acne and impetigo infections.
Hibitane (ICI)	Chlorhexidine	1,6-di(4-chlorophenyl-diguanido) hexane	Parachloraniline + hexamethylene diamine intermediates → chlorhexidine	(b)	Small animals Hibitane acetate, 2000 mg/kg Hibitane gluconate, 1800 mg/kg	Bactericide effective against gram-positive and gram-negative bacteria. Disinfectant for skin, wounds and burns, surgical gloves and operating theaters. Treatment of exzema and impetigo skin infections.

(a)

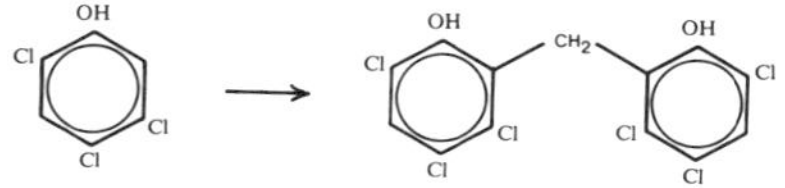

[Structure: 2,4,5-Trichlorophenol converted to 2,2'-dihydroxy-3,3',5,5',6,6'-hexachlorodiphenylmethane — two chlorinated phenol rings joined by CH$_2$]

(b)

[Structure: Cl—C$_6$H$_4$—NH—C(=NH)—NH—C(=NH)—NH—(CH$_2$)$_6$—NH—C(=NH)—NH—C(=NH)—NH—C$_6$H$_4$—Cl]

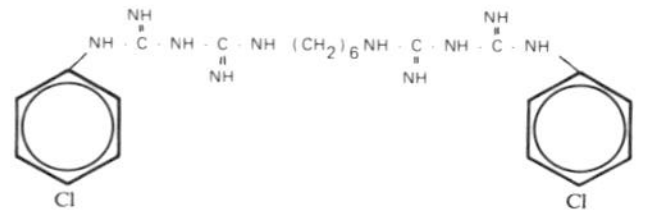

Hexachlorophene, on the other hand, long established as a bacteriacide, has frequent side effects including diarrhea, abdominal pain and distension, dizziness, headache, muscular weakness, and drowsiness, which have followed the administration of repeated doses of 20 mg/kg body weight. Photosensitivity and skin sensitization have also occurred under similar conditions.[3] A full list of the products containing chlorhexidene and hexachlorophene is given in the Martindale Pharmacopoeia.[2,3]

It is on the question of the cleansing of the newly born infant, however, that there is a difference of opinion as to which of the two bacteriacides is the most effective. Maternity clinics and nurseries are particularly open to bacterial cross infection by the staphylococcal organisms. One of the most common is *Staphylococcus aurens*. In the 1940s this organism was responsible for frequent epidemics in nurseries with a consequent increase in infant mortality. When hexachlorophene was first marketed by Givaudan in the late 1940s it proved to be effective both in the routine containment of *S. aurens* and in controlling the organism in the case of an epidemic. As a result of its efficiency, hexachlorophene rapidly replaced the bacteriacide "Triple Dye" in use at that time, to become the most widely used antibacterial agent in nurseries.

Two events in 1971 and 1972, however, caused users to reconsider their judgment. The first was a report by the U.S. Environmental Protection Agency that research[4,5] commissioned had shown that hexachlorophene had caused cerebral edema, brain lesions, and hindlimb paralysis in rats. Shortly thereafter on 8 December 1971 the U.S. Food and Drug Administration recommended that the use of hexachlorophene in nurseries be limited.[6] The second event was the death in France of 36 infants following the use of talcum powder containing 6% hexachlorophene. The neurological damage which led to the death of the children was caused by a 20-fold increase of hexachlorophene concentration in the talc, the result of a manufacturing error.

It took eight years for the case to be brought to court, seven of which were spent in investigating the background to the case. In the trial, which began in the first week of October 1979, six defendants stood accused of causing the deaths of the children. The accused were from the two French companies, Morhange and Setico, which were directly involved with the tragedy. The talc was sold under the Morhange trademark but prepared by Setico. When the case started both companies were insolvent. In view of this the prosecution directed their attack against the Geneva-based company Givaudan, the manufacturer and supplier of the hexachlorophene for Setico.

The details of the case are tragic. In 1972, an outbreak of rashes and sores affected babies in northeastern France. Mothers reacted by applying talcum powder to dry the skin. But far from containing the problem, the outbreak spread only to be followed by the infants having convulsions and lapsing into coma, attributable to the absorption of hexachlorophene through the skin. When Morhange talc was finally identified as the factor common to all these symptoms, 36 children had died, 8 had been permanently maimed, and some 15 others had serious health problems.[7] The number of children injured was much higher than this, however. According to Givaudan 145 children had extremely serious nerve damage as a result of their exposure to the talc.[8]

One of the accused in the case was Dominique Civel. An employee of Setico, Civel testified in court that he had only been given one day's training in the mixing of the talcum powder, which should never have contained hexachlorophene at all.[8] The Morhange talc was prepared in a room along with a second preparation, which required hexachlorophene. Civel's instructions were to keep the hexachlorophene on the floor below the powders he was mixing and to bring the exact quantity of hexachlorophene up by bucket as required. He decided, however, that it would be neater to have the hexachlorophene barrel arranged alongside the others he was using and moved it on his first day. But not only was Civel new to his job, he was often off-duty and a replacement was provided. The prosecution alleged in court[7] that it was Setico's policy to tip any powder swept off the floor into the talcum powder. One day 39 kilos of hexachlorophene powder was on the floor and it was claimed that Civel's replacement mistook the hexachlorophene for floor sweepings and tipped the bacteriacide into the talc.

As the only solvent company involved in the episode, Givaudan was charged in court with not making its clients sufficiently aware of the danger of hexachlorophene. The company denied the charge but felt morally bound to provide compensation. Most parents who lost a child accepted out-of-court settlements of £2800. Half the families who had to bring up a handicapped child settled for £160,000 each. The other half together with some of the bereaved parents believed these payments to be inadequate and charged Givaudan accordingly. Observers believed that the parents faced an uphill task and that the court could well rule that Givaudan had no case to answer[7] and that it was not even morally obliged to compensate the victims.

The two-month trial ended in December 1979; the verdicts were handed down on 11 February 1980. Six people were charged with involuntary homicide and injury in connection with the marketing of the baby powder. Those charged included senior management staff of Morhange and Setico, two Setico employees, and Hubert Flahault, President of Givaudan—France, the manufacturer of the hexachlorophene. All were found guilty as charged with the exception of Dominique Civel, who was acquitted. Sentences for those found guilty, ranging from 1 to 20 months, were all suspended. The court ordered Setico to pay more than £250,000 in costs and

£75,000 in damages and interest to the 45 civil plaintiffs, as well as £5000 to the parents of each child. The five defendants found guilty all plan to appeal.[9]

Following the reports in 1972 of the tragedy in France, many maternity units in the United Kingdom reduced the amount of hexachlorophene used for infant washes. Currently less than half use hexachlorophene at all. Others rely on alcohol, used either alone or with chlorhexidine. Consultants at one cross-infection laboratory, Colindale Public Health Laboratory in London, now recommend nurseries to avoid the use of hexachlorophene altogether for routine washes.[10] The reasoning behind this recommendation lies in the evolution of the bacteria *S. aurens*, which hexachlorophene is known to kill.

The bacterium has, since the 1940s, evolved through several different forms and is now active as a complex of *S. aurens*. According to Colindale's consultants the earlier strains of *S. aurens*—strains 80/81—had the capacity to create epidemics; however, the current strains are far less virulent and are no longer causing outbreaks in British hospitals which result in infant deaths. In the absence of an epidemic the consultants do not feel justified in recommending that hexachlorophene be used; its use is considered to present a greater threat than that represented by the staphylococcus itself.

One irony which the Colindale scientists report is that many maternity clinics use a concentration of hexachlorophene to kill *S. aurens* which is too low to kill the bacterium. This fact was discovered in a survey by the scientists of some 60 maternity clinics in the wake of the hexachlorophene-linked deaths in France. The survey revealed that a third of the clinics then using a fairly low concentration of hexachlorophene as a bacteriacide reverted to an even less concentrated preparation. This latter formulation, say the Colindale scientists, will not kill *S. aurens*.

Many in the medical profession would like hexachlorophene to be available at all times. Hexachlorophene, they argue, is invaluable in controlling staphylococcal infections in the nursery.[11-14] But one editorial writer in the *British Medical Journal* was not so sure. In the editorial on 5 February 1977,[15] he wrote: "For hexachlorophene our information is inadequate for scientific support either of retention or removal of the restrictions on its use. Before a drug is administered to man we should be able to show a need for it and to relate the safety of treatment to the hazard of the disease. We lack epidemiological evidence of continued need for HCP (hexachlorophene) in the nursery, and there is insufficient information to establish a safe level of human exposure." Arguing the case for more information on the absorption, metabolism, effect, and pharmacokinetics of the bacteriacide in animals and humans the author concludes: "Without such information HCP may well become another of these victims of the certainty with which a regulatory authority will ban a compound on incomplete evidence and the scientific uncertainty inevitable with negative findings in clinical practice."

The same plea was made five years prior to this at a conference on hexachlorophene. Acknowledging that the bacteriacide was effective in the control of *S. aurens*, the author, Louis Gluck of the School of Medicine at the University of California, called for more information on hexachlorophene and ". . . particularly more about the metabolism in the human, absorption, excretion, body storage, tissue distribution and so forth."[16] It seems that little has been done in the intervening years between this 1972 appeal and the *British Medical Journal* editorial.

Fears expressed in the editorial about regulatory authorities acting to control hexachlorophene have become a reality in Sweden. Products containing hexachlorophene in Sweden can now only be prescribed by physicians; in the past they were sold over the counter without any restrictions.

The Swedish Medical Authorities proposed this measure in 1979 and it became law in 1980. The reason for their decision was a report in 1978 claiming that pregnant mothers exposed to hexachlorophene soaps had given birth to a higher proportion of severely malformed children than mothers not so exposed.[17] Dr. Hildegard Halling of the Department of Chronic Somatic Diseases at Sweden's Södertälje Hospital reported her findings in full at a conference in New York in June 1978. Halling said that retrospective studies performed in six Swedish hospitals showed that out of 460 children born to mothers who used hexachlorophene soaps or handcream 10–60 times a day in the first trimester of pregnancy, 25 were born severely malformed. No severe malformations were noted among the 233 children born to the control group of mothers not using the soaps or creams. The incidence of minor malformations was also higher in the hexachlorophene-exposed group.

This Halling study—which reports a wide range in the type of abnormalities seen—is, as yet, unconfirmed. A team from the U.S. Food and Drug Administration has studied the findings in detail and concluded that Halling had not proved the case against hexachlorophene. The Swedish Medical Board expressed a more cautious opinion. The Board had been unable to substantiate a previous report of Halling's in 1977 in which she suggested that hexachlorophene was responsible for an increase in the rate of malformation in children born in Swedish hospitals[18]; it noted, however, that there was a high rate of malformations in the Gothenberg area—one of those studied by Halling—but could offer no explanation for this.[19]

As for Halling's later report in 1978, the Swedish Medical Board could not rule out hexachlorophene as the cause of the high incidence of malformations. They noted that the consumption of hexachlorophene had not been taken into account by Halling and this they thought was a serious omission in her study. But they considered that she had made a valuable observation. As for hexachlorophene the Board pointed out that it had been shown to cross the placenta and could put the developing fetus at risk. To prevent this from occurring a restriction on hexachlorophene use was proposed, and enforced in 1980.[20]

A restriction on the product is one thing, but could hexachlorophene be replaced altogether?

A replacement for hexachlorophene has not been tested under all circumstances and this is the main reason for the reticence of the medical profession in permitting its withdrawal. In the routine control of *S. aurens*, chlorhexidene is as effective as hexachlorophene according to the Colindale scientists and others.[21-23] Chlorhexidene is also reported to be just as efficient in controlling local epidemics of the present milder strain of the bacterium.[10,24] It has not been tested, however, with more lethal strains of the staphylococcus, whereas hexachlorophene has been shown to control dangerous epidemics of *S. aurens* in the past. Some bacteriologists anticipate that chlorhexidene—only recently approved for sale in the United States—will prove to be equally effective, but are reluctant to recommend it unequivocally as an alternative until this has been demonstrated conclusively.

References

1. Lowbury, E. J. L., Lilly, H. A., and Bull, J. P. Disinfection of hands: Removal of transient organisms, *British Medical Journal II*, 230-232 (1964).
2. Martindale, *The Extra Pharmacopoeia*, 7th ed., The Pharmaceutical Press, London (1977), Chlorhexidene, p. 508.
3. Martindale, *The Extra Pharmacopoeia*, 7th ed., The Pharmaceutical Press, London (1977), Hexachlorophene, p. 522.
4. Kimbrough, R. D. Review of the toxicity of hexachlorophene, *Archives of Environmental Health* 23, 119 (1971).
5. Kimbrough, R. D., and Gaines, R. D. Hexachlorophene effects on the rat brain. Study of high doses by light and electron microscopy, *Archives of Environmental Health* 23, 114-118 (1971).
6. Food and Drug Administration: Drug Bulletin, Health Service and Mental Health Administration, U.S. Department of Health, Education and Welfare, (December 1971).
7. Smyth, Robin. *The Observer*, 7 October 1979.
8. Givaudan statement, Morhange Talcum Power—France, July 1978.
9. Anonymous. *The Guardian*, 12 February 1980.
10. Hay, Alastair. Seveso: The Aftermath, *Nature (London)* 263, 538-540 (1976).
11. Gezon, H. M., Thompson, D. J., Rogers, K. D., Hatch, T. F., Rycheck, R. R., and Yee, R. Control of staphylococcal infections and disease in the newborn through the use of hexachlorophene bathing, *Pediatrics* 51, 2(II), 331-344 (1973).
12. Pleuckhahn, V. D. Hexachlorophene and the control of staphylococcal sepsis in a maternity unit in Geelong, Australia, *Pediatrics* 51, 2(II), 368-382 (1973).

13. Gezon, H. M., Schaberg, M. J., and Klein, J. O. Concurrent epidemics of staphylococcus aurens and Group A steptococcus disease in a newborn nursery—control with penicillin G and hexachlorophene bathing, *Pediatrics* **51,** 2(II), 382–394 (1973).
14. Corner, B. D., Alder, G., Burman, D., and Gillespie, W. A. Hexachlorophene—Yes or No?, *British Medical Journal* **1,** 636 (1977).
15. Editorial. Hexachlorophene—Yes or No?, *British Medical Journal* **1,** 337 (1977).
16. Gluck, L. A perspective on hexachlorophene, *Pediatrics* **51**(II), 400–412 (1973).
17. Halling, H. Suspected link between exposure to hexachlorophene and malformed infants, *Annals of the New York Academy of Sciences* **320,** 426–435 (1979).
18. Halling, H. Misstänkt samband mellan hexaklorofenexposition och missbild ningsbörd, *Swedish Medical Journal* **74,** 542–546 (1977).
19. Westerholm, B. Kommentar, *Swedish Medical Journal* **74,** 545 (1977).
20. Larssen, Sune. Personal communication, 3 September 1980.
21. Maloney, M. H. Chlorhexidene a hexachlorophene substitute in the nursery, *Nursing Times*, 11 September 1975.
22. Tuke, W. Hibiscrub in the control of staphylococcal infection in neonates, *Nursing Times*, 11 September 1975.
23. Geoge, R. H. The effect of antimicrobial agents in the umbilical cord, *Chemotherapy* **3,** 415–420 (1976).
24. Senior, N. Some observations on the formulation and properties of chlorhexidene, presented at Symposium on Skin-Environmental Responses and Protection organized by the Society of Cosmetic Chemists of Great Britain, Oxford, United Kingdom, 11 April 1972.

4

2,4,5-T, Trichlorophenol, Pentachlorophenol, and Polychlorinated Dibenzofurans

The current advocacy of the merits of 2,4,5-T by representatives of the chemical industry and government departments responsible for agriculture would carry more weight if the herbicide was unique, which it is not. 2,4,5-T was developed in the wake of the identification of the plant growth hormones, auxins. It was the observation that high concentrations of the growth hormones were toxic to plants which led to the development of synthetic acids for weedkillers of which 2,4,5-T is one. The herbicide is an effective killer of broad-leafed plants but relatively innocuous where cereals and grasses are concerned.

Although 2,4,5-T is no longer manufactured in the United Kingdom, the herbicide is still imported and available for sale in many gardening shops. The U.K. Ministry of Agriculture, Fisheries and Food in its handbook *Approved Products for Farmers and Growers*[1] lists 15 preparations of 2,4,5-T currently on sale. The handbook says that 2,4,5-T can be used for the "control of woody weeds in a variety of situations and can be used selectively in conifers." Caution when using 2,4,5-T is recommended by the handbook, which states that "spray drift" should be avoided. Farm stock it advises should be kept away from sprayed areas for at least two weeks. This last recommendation is to ensure that all weeds—particularly poisonous ones—are given sufficient time to die. Allowing animals to graze before this could lead to fatalities if poisonous plants are consumed. The herbicide itself is not considered to be life threatening for the animals. However, 2,4,5-T is "harmful to fish," the booklet says.

There is a second array of 2,4,5-T preparations available in the United Kingdom. These have been cleared for use by MAFF, but not approved by the Ministry. The distinction is more than one of semantics. Chemicals which are approved are subject to more rigorous examination by MAFF's Pesticides Advisory Committee. Similar schemes to vet herbicides and pesticides before they are allowed to be marketed exist in other countries but the detailed procedures are often different. This would account for some countries banning 2,4,5-T, but others—such as Britain—allowing it to remain in use.

The preparations of 2,4,5-T which are available in the United Kingdom and other countries, consist, in the main, of three types. These are ester formulations, unformulated esters, and mixtures of 2,4,5-T with other herbicides such as 2,4-D (2,4-dichlorophenoxyacetic acid).

In ester formulations the acetic acid side chain of 2,4,5-T is made to react with various alcohols such as ethanol, propanol, and butanol to form an ester and make a longer-acting herbicide. The ester formulations have a bulkier side chain attached to the parent benzene ring and this slows down the rate of 2,4,5-T breakdown.[1]

Unformulated esters are usually diluted in oil. The major advantage with preparations of the herbicide in oil is that evaporation, a problem with water-based formulations, is considerably reduced. Another advantage with oil formulations is their use in ultralow volume sprayers. This technique—used frequently in forestry—allows very fine droplets (almost a mist) to be spread over a wide area. This, in turn, means a more efficient use of 2,4,5-T and thus a reduction in the amount of herbicide required to kill vegetation.

Many articles on the effectiveness, mode of action, toxicity, and ultimate fate of 2,4,5-T in agriculture and silviculture (forestry) have been published since the herbicides introduction as a commercial proposition. The number of scientific publications must run into the thousands and only the briefest of references can be made to them here. One of the more recent reviews of 2,4,5-T was published by the Swedish Natural Research Council, NFR, in 1978. Called *Chlorinated Phenoxy Acids and Their Dioxins*, the book, a compilation of scientific papers presented at a Swedish symposium on the subject in February 1977, discusses the mode of action, health risks, and environmental effects of 2,4,5-T and its related phenoxy acids.[3]

Similar information is available on the alternatives to 2,4,5-T, of which there are three: "Amcide," "Glyphosate," and "Krenite"; all three are effective brushwood killers. Amcide has been available in the United Kingdom since the early 1960s; Glyphosate was a later edition and Krenite is the most recent.

These alternatives, when compared with 2,4,5-T, have some definite advantages. Amcide, the trade name for ammonium sulfamate, marketed by Albright and Wilson, has an LD_{50} value for rats of 3900 mg kg^{-1}; Glyphosate, made by Monsanto, has an LD_{50} value of 4900 mg kg^{-1}. These toxicity ratings are well below that for 2,4,5-T (LD_{50} value 300 mg kg^{-1}).

Krenite, made by DuPont, has an LD_{50} value of 24,000 mg kg^{-1}, a toxicity rating 1/80 of that for 2,4,5-T.

It is conceivable that the lethal dose which will kill half the animals in a test group (the LD_{50} value) may be too low for some of the 2,4,5-T preparations currently available. Earlier preparations of the herbicide may have had higher concentrations of the toxic dioxin contaminant than the current 0.1 ppm required by law since the early 1970s. In Britain, dioxin levels must now be kept below 0.01 ppm.

Nevertheless, the reactions to produce all three 2,4,5-T alternatives present none of the potential hazards associated with trichlorophenol manufacture in terms of toxic by-products.[4-6] Details about the manufacturing processes, where available, are given in Table IX. Neither the Monsanto nor DuPont corporations were prepared to divulge this information, both insisting that it remained confidential. A DuPont spokesman was insistent, however, that as far as the U.S. Environmental Protection Agency (EPA) was concerned, there is not the slightest doubt about the safety and use of Krenite. Indeed, in the case of Krenite, the EPA has declared that in view of its comparatively low toxicity it is safe for use even on land adjacent to domestic water supply reservoirs and streams.[7]

The disadvantages associated with the three alternatives are primarily those of cost. Albright and Wilson, the original manufacturers of ammonium sulfamate, now only market the chemical for the Japanese company Nissan under the Amcide trade name. According to a spokesman for Albright and Wilson, the Japanese company developed a more economic process for the manufacture of ammonium sulfamate. In the event, it became necessary for the British company to cease production and operate as a conduit for the Japanese product.

Both Amcide and Glyphosate are more expensive than 2,4,5-T when the concentrations of herbicide necessary to kill the same plant ratio are considered. The U.K. Forestry Commission has done a good deal of research on these alternatives. According to the Commission's Research Department, when plant kill ratios are taken into account, Glyphosate is 4–5 times as costly as 2,4,5-T.[7] The Forestry Commission spokesman said that they had not used Krenite in trials but that it would almost certainly prove to be an expensive product. As for Amcide, a second spokesman claimed that as far as use in forestry was concerned this product was not a useful alternative to 2,4,5-T.

In contrast to the phenoxy acid which is absorbed through leaves following spraying, Amcide must be applied to cut surfaces on plants. Its method of application is therefore more labor intensive. As far as Glyphosate is concerned, its cost disadvantage is the only real drawback. It certainly is an alternative to 2,4,5-T in forestry. In 1973 Norwegian authorities

TABLE IX. Herbicides

Product name	Common name	Chemical name	Method of production
2,4,5-T (Dow Chemical Co. U.S.)	2,4,5-T	2,4,5-Trichlorophenoxyacetic acid	Benzene ⟶ tetrachlorobenzene ⟶ trichlorophenol ⟶ 2,4,5-T ↘ TCDD
Amicide (Nissan, Japan)	Ammonium sulfamate	Ammonium aminosulfonate	Sulphur trioxide + ammonia ↓ sulfonic acid ↓ ammonium sulfamate $SO_3 + NH_3 \longrightarrow NH_2SO_3H \longrightarrow NH_2SO_3NH_4$
Roundup (Monsanto U.S. and U.K.)	Glyphosate	N(phosphonomethyl) glycine	Information not made available
Krenite (Dupont U.S.)	Not yet finalized	Ammonium ethyl carbomylphosphonate	Information not made available
2,4-D (Hedonal)	2,4-D	2,4-Dichlorophenoxyacetic acid	2,4-Dichlorophenol + monochloroacetic acid in aqueous NaOH

banned the use of 2,4,5-T. Faced with the need to find a useful alternative Norwegian foresters experimented with a number of herbicides, of which Glyphosate was one. Their conclusions were[8] that Glyphosate was indeed an alternative to 2,4,5-T and that it may in fact be more longer acting than 2,4,5-T and so obviate the need for respraying the following year.

From the above, it is clear that alternatives to 2,4,5-T do exist. Were the phenoxy acid to be withdrawn from use through fears about its safety, current users of the product would still be able to remove unwanted vegetation with the available alternatives.

Structure	Toxicity	Properties
2,4-D structure: benzene with OCH₂COOH and two Cl (ortho, para) — OCH$_2$COOH, Cl, Cl	LD$_{50}$: acute oral Rats: 300 mg/kg	Selective herbicide. Control of woody plants. Absorbed through leaves. Following spraying at least 14 weeks required before replanting.
$H_2N-\overset{O}{\underset{O}{\overset{\|}{\underset{\|}{S}}}}-ONH_4$ (NH$_2$SO$_3$NH$_4$)	Rats: 3900 mg/kg	Nonselective control of wide range of vegetation. Absorbed through cut surfaces. Replanting 6–8 weeks following application.
$HO-\overset{O}{\overset{\|}{C}}-CH_2-\underset{H}{N}-CH_2-\overset{O}{\overset{\|}{P}}-OH$, with OH on P	Rats: 4900 mg/kg	Foliar acting herbicide for systemic control of annual and perennial grasses and broad-leaved weeds. Replanting 2–3 weeks following application.
$CH_3CH_2O-P(\uparrow O)(O^-)-\overset{O}{\overset{\|}{C}}-NH_2 \cdot NH_4^+$	Rats: 24,000 mg/kg Guinea pigs: 7380 mg/kg Mallard ducks: 10,000 mg/kg	Brush control agent for wide range of plant species. Absorbed through foliage, buds, and stems. Rapidly decomposed in soils.
2,4,5-T structure: benzene with OCH$_2$COOH and three Cl (2,4,5 positions)	LD$_{50}$: acute oral Rats: 1500 mg/kg Mice: 375 mg/kg	Weed control. Increase latex output of old rubber trees. Selective herbicide.

2,4,5-Trichlorophenol

Although the principle purpose behind 2,4,5-trichlorophenol production is the synthesis of the herbicide 2,4,5-T and the bacteriacide hexachlorophene, the chlorinated phenol is still used to some extent as a slime control agent in the paper-making industry (see Table X). It has never been as popular as pentachlorophenol, used for the same purpose, and what use it has had has been much reduced, owing mainly to the criticisms of the unrestricted industrial use of polychlorinated biphenyls (PCBs) as well as DDT, advanced by the environmental lobby.

TABLE X. Slimicides

Chemical	Method of production
2,4,5-Trichlorophenol	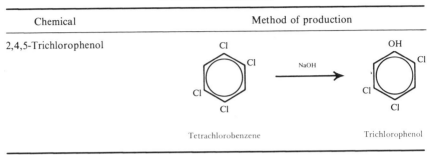 Tetrachlorobenzene → (NaOH) → Trichlorophenol

The Port of London Authority (PLA) was instrumental in controlling effluent discharged directly into the Thames. According to a spokesman for Bowater Packaging Corporation of Gravesend, the PLA forced every factory discharging effluent directly into the Thames to disclose details of the chemicals it was using in paper processing. As a result of this action and the criticisms of some of the chemicals then in use, the spokesman said that the whole industry had tended to "go away from products which are harmful to the environment."

According to the Printing Industries Research Association (PIRA) each paper mill is a law unto itself, and consequently each chooses what it considers to be the best method of controlling fungal growth with slimicides. Chemicals commonly used for this purpose according to PIRA now include methylene bis (thiocyanate), organobromine, and organosulfur compounds.[7] Mercury-based compounds in the parts-per-million range are considered to be safer than other chemicals in higher concentrations.

Pentachlorophenol

Pentachlorophenol is one of the major chlorinated phenols in use. Like 2,4,5-trichlorophenol, it usually contains contaminating dioxins.

First produced in the 1930s, pentachlorophenol is used widely for the control of mold and mildew growth on wood, as a pesticide for controlling termites,[9] as a molluscicide in the control of snails in the schistosomiasis life cycle,[10] and as a herbicide in pineapple and sugarcane fields.[11,12]

Mammals are said to metabolize pure pentachlorophenol readily and to excrete all traces of the compound.[9] However, there is a hazard associated with the chemical and this involves the presence of chlorinated dibenzodioxin contaminants. The principal dioxin isomers present in pentachlorophenol are hexachlorodibenzo-*p*-dioxins, heptachlorodibenzo-*p*-

Toxicity	Uses
LD_{50}: oral Rats: 2960 mg/kg	Prevent slime growth in paper-making industry. Now largely superceded.

dioxins, and octachlorodibenzo-p-dioxin. The content of dioxin impurities varies between different batches of the chemical. One technical grade batch of the chemical is reported to contain 8 ppm hexa-, 520 ppm hepta-, and 1.38 ppm octachlorodibenzo-p-dioxin, as well as a selection of chlorinated dibenzofurans[13] (see Chapter 1 for more details on the chemistry of these compounds). The maximum concentrations of these contaminants is generally not thought to exceed 100 ppm hexa-, 200 ppm hepta-, and 2.0 ppm octachlorodibenzo-p-dioxin.[9] The most toxic dioxin isomer 2,3,7,8-tetrachlorodibenzo-p-dioxin, produced in the manufacture of 2,4,5-trichlorophenol, has never been reported to occur in pentachlorophenol.[11] Its presence would not be expected in view of the method of synthesizing pentachlorophenol.

Pentachlorophenol metabolism will not be reviewed here. Readers interested in the subject should refer to the comprehensive reviews already available.[9,14]

It is worth noting, however, that pentachlorophenol contaminated with dioxin isomers containing six, seven, and eight chlorine atoms can have drastic effects. This was first evident in 1957 following the deaths of millions of broiler chickens in the eastern and midwestern parts of the United States. The immediate cause of death was fluid retention in the pericardium (the sac surrounding the heart) and the abdomen—referred to as chick edema disease. Investigations by the U.S. Food and Drug Administration soon established that the common factor in all the deaths was a low-cost chick feed. The feed had been supplemented by fats obtained from tallow, which had, in turn, been rendered from cattle hides previously treated for the purposes of preservation with pentachlorophenol.[15] Once the fat was eliminated from the chick feed, the problem was solved as far as the broiler owners were concerned. But this was not the end of the story. It was not pentachlorophenol itself which had caused the edema, but 1,2,3,7,8,9-hexachlorodibenzo-p-dioxin, which was subsequently identified in the toxic fat in 1966.[16] Experiments with this dioxin showed unequivocally that it would cause edema in chickens.

Contaminated pentachlorophenol is also reported to have been the cause of ill health and death in cattle in the United States. The animals were thought to have chewed and rubbed against wood in a newly constructed barn, the timber of which had been treated with pentachlorophenol.[17] Research by a team under the direction of Dr. Eugene McConnell, head of the Comparative Pathology Section of the National Institute of Environmental Health Sciences, has shown that contaminated pentachlorophenol will cause medical problems in cattle which are not observed with comparable doses of noncontaminated material. Progressive anemia, decreasing body weight and feed efficiency, and lesions of the bladder were observed in heifers receiving the contaminated phenol. The severity of the effects increased with increasing doses of the chemical. McConnell attributes the effects to contamination of pentachlorophenol by dibenzo-*p*-dioxins or dibenzofurans.[17]

Human health problems associated with exposure to pentachlorophenol have received more attention only in recent years. With hexachlorodibenzo-*p*-dioxins now known to be carcinogenic in animals,[18] the subject is one under intense scrutiny by regulatory authorities in the United States and the United Kingdom.

The U.S. Environmental Protection Agency is reported to be investigating allegations that residents near the Lexington Blue Grass army depot in Kentucky have a high incidence of leukemia, which they claim may be due to exposure to pentachlorophenol used to preserve ammunition boxes at the depot. The wood from the boxes was sold to residents for the construction of barns, chicken coops, pigpens, or as firewood. According to the EPA, a preliminary investigation confirms that there is a high incidence of leukemia in the area. A major epidemiological study is being mounted by the National Institute of Occupational Safety and Health to confirm this finding and to try and establish the cause of the disease.

In the United Kingdom the Health and Safety Executive (HSE) has investigated workers exposed to chlorinated dibenzo-*p*-dioxins at Monsanto's pentachlorophenol plant at Newport, South Wales.[19] The plant was closed in 1979 after requests from the HSE that extra safety features be incorporated to reduce worker exposure to the chemical were considered by the company to be too expensive. Men formerly employed on the site and exposed to pentachlorophenol are reported to have the skin disease chloracne and a variety of blood fat disorders. In particular the men are reported to have raised serum triglyceride levels and some elevation of blood cholesterol. This group of workers is still under observation and is likely to be monitored over a period of years. (More information about these men is given in Chapter 6: Other Incidents).

Polychlorinated Dibenzofurans

In Chapter 1 reference was made to 1684 people in Japan who consumed rice oil contaminated with polychlorinated biphenyls (PCBs). The PCBs were in turn contaminated with polychlorinated dibenzofurans (furans). The toxicity of the furans is similar to that of the dioxins.

As the contaminated rice oil was directly responsible for the health problems of the 1684 people the group became known as the Yusho victims. (Yusho is Japanese for oil disease.) The history of their exposure to PCBs and some of the initial medical studies are well documented in Kentaro Higuchi's book *PCB Poisoning and Pollution*.[20] Many doctors believe that the severity of the medical disorders experienced by the Yusho patients cannot be explained solely by their exposure to PCBs. The contaminating toxic furans are thought to have made the problems a good deal worse.

Yearly follow-up studies of the patients are made and in the latest report[21] reference is made to the skin lesions every victim developed. The authors report that "As a whole, skin symptoms have diminished gradually, while, on the other hand, continual subcutaneous cyst formation with secondary infection are still occurring in a relatively small number of the severe grade patients." Depressingly, the authors conclude: "We have no essentially effective treatment for the skin changes of Yusho and only some consolatory prescriptions of drugs or surgical treatment are available."

Mortality studies on Yusho patients do not conclusively link death with PCB or furan poisoning. It is known, however, that certain patients died suddenly in the early stages of poisoning. Their symptoms were similar to those seen in adrenal crisis, suggesting that PCBs might interfere with the endocrine system either directly or indirectly.[22] The symptoms reported include severe pain in the abdomen, nausea, vomiting, and lower arterial blood pressure associated with collapse of the peripheral blood vessels.[22]

Disturbances of the menstrual cycle, the cycle being either prolonged, shortened, or irregular, have been reported for many of the female Yusho patients. Respiratory distress with coughing and sputum production was yet another common complaint. PCBs after ingestion become localized to some extent in the lungs and bronchial tract, are excreted via the respiratory tract, and would simultaneously cause respiratory distress.[22] Blood fats were also elevated by the PCB poisoning, with high serum triglycerides a common finding. Cholesterol, on the other hand, was reported to be unaffected,[23] and blood triglyceride levels eventually returned to the recognized normal ranges.[24]

One of the more disturbing aspects of the PCB poisoning was the effect on fetal and infant life. Birth abnormalities and still births have been reported in the children born to pregnant women with PCB poisoning.[25,26] Whether the number of cases reported is higher than the norm is not, however, clear. What is clear is the low birth weights recorded in most cases for babies born to poisoned mothers.[26] Many of those infants faced a double jeopardy. Not only were they small-for-date, but breast feeding brought on characteristic signs of PCB poisoning presumably due to PCB contamination of the breast milk. Pigmentation of the skin, tooth and bone defects, and growth retardation were additional complications for these infants.[27]

References

1. *Approved Products for Farmers and Growers*, published annually by U.K. Ministry of Agriculture, Fisheries and Food.
2. Leng, M. L. ACS Symposium on Fate of Pesticides in Large Animals, paper No. 73 in Division of Pesticide Chemistry, Centennial Meeting of the American Chemical Society, San Francisco, 29 August–3 September 1976).
3. *Chlorinated Phenoxy Acids and their Dioxins*, Ecological Bulletins, No. 27, ed. C. Ramel, published by Swedish Natural Science Research Council, NFR, Box 23136, S-104 35, Stockholm, Sweden (1978).
4. Hawkins, R. E. C. Technical Department, Organics Sector, Albright and Wilson, personal letter, 10 September 1976.
5. Johnson, E. Agricultural Division, Monsanto. Personal letters, 24 September 1976 and 26 October 1976.
6. Hay, S. J. B. Biochemicals Department, Dupont (UK) Ltd. Personal letter, 30 September 1976.
7. Hay, A. Seveso: the aftermath. *Nature (London)* **263**, 538–540 (1976).
8. Lund-Hoie, K. N-phosphonomethylglycine (glyphosate) an alternative to commercial pre- and post-emergence herbicides for the control of unwanted plant species in forest plantations in Norway. *Meldinger fra Norges Landbrukshogskole* (Scientific Reports of the Agricultural University of Norway) **54**, 1–14 (1975).

Pentachlorophenol

9. Ahlborg, U. G. Metabolism of chlorophenols: studies on dechlorination in mammals, thesis, Rapport fran statens naturvardoverk (1977).
10. Roushdy, M . Z., El-Gindy, M. S., Ayad, N., Mousa, A. H. Effect of sub-lethal doses of molluscicides on the susceptibility of *Bulinus truncatus* snails to infection with Schistosoma haematobium, *Egyptian Journal of Bilharziasis* **1**, 85–90 (1974).
11. Bevenue, A., and Beckman, H. Pentachlorophenol. A discussion of its properties and its occurrence as a residue in human and animal tissues, *Residue Revue* **19**, 90–134 (1967).
12. Bevenue, A., Haley, T. J., and Klemmer, H. W. A note on the effects of a temporary exposure of an individual to pentachlorophenol, *Bulletin of Environmental Contamination and Toxicology* **2**, 293–296 (1967a).
13. Goldstein, J. A., Friesen, M., Linder, R. E., Hickman, P., Hass, R. J., and Bergman, H. Effects of pentachlorophenol on hepatic drug-metabolizing enzymes and porphyria related to contamination with chlorinated dibenzo-*p*-dioxins and dibenzofurans, *Biochemical Pharmacology* **26**, 1549–1557 (1977).
14. Ranga Rao, K., ed. *Pentachlorophenol—Chemistry, Pharmacology and Environmental Toxicology*, Plenum Press, New York (1978).

15. Firestone, D. Etiology of chick edema disease, *Environmental Health Perspectives* **5**, 59-66 (1973).
16. Cantrell, J. S., Webb, N. C., and Mabis, A. J. Identification and crystal structure of a hydropercardium-producing factor: 1,2,3,7,8,9-hexachlorodibenzo-*p*-dioxin, *Acta Crystalography* **B25**(I), 150 (1969); see also *Chemical and Engineering News* **45**, 10 (1967).
17. Dickson, D. PCP dioxins found to pose health risks, *Nature (London)* **283**, 418 (1980).
18. Hay, A. Halogenated hydrocarbon effects, *Nature (London)* **274**, 533-534 (1978).
19. Anonymous. Notes of informal meeting to discuss the problem of chloracne. U.K. Health and Safety Executive, 7 November 1979.

Polychlorinated Dibenzofurans

20. Higuchi, K. *PCB Poisoning and Pollution*, Kodansha Limited, Academic Press, New York (1976).
21. Asahi, M., Koda, H., Urabe, H., and Toshitani, S. Dermatological Symptoms of Yusho: Alterations in this decade, *Fukuoka Acta Medica* **70**(4), 172-180 (1979).
22. Hirayama, C. Clinical aspects of PCB poisoning, in *PCB Poisoning and Pollution*, ed. K. Higuchi, Kodansha Limited, Academic Press, New York (1975), pp. 87-104.
23. Uzawa, H., Ito, Y., Notomi, A., and Katsuki, S. Hyperglyceridemia resulting from intake of rice oil contaminated with chlorinated biphenyls, *Fukuoka Acta Med.* **60**, 449-454 (1969).
24. Okumura, M., Yamanaka, M., and Nakamuta, S. Ten year follow-up study on serum tricycleride levels in 24 patients with PCB poisoning, *Fukuoka Acta Med.* **70**(4), 208-210 (1979).
25. Taki, I., Hisanga, S., and Amagase, Y. Report on Yusho (chlorobiphenyls poisoning) in pregnant women and their fetuses. *Fukuoka Acta Med.* **60**, 471-474 (1969).
26. Funatsu, I., Yamashita, F., Yoshikane, T., Funatsu, T., Ito, Y., Tsugawa, S., Hayashi, M., Kato, T., Takushiji, M., Okamoto, G., Asima, A., Adachi, N., Takahashi, K., Miyahara, M., Tashiro, Y., Shimomura, M., Yamasaki, S., Arima, T., Kuno, T., Idi, H., and Ide, I. Chlorobiphenyl induced fetopathy, *Fukuoka Acta Med.* **62**, 139-149 (1971).
27. Yamaguchi, A., Yoshimura, T., and Kuratsune, M., A survey on pregnant women having consumed rice oil contaminated with chlorobiphenyls and their babies, *Fukuoka Acta Med.* **62**, 117-122 (1971).

5

Chloracne

One of the earliest and most obvious symptoms of human exposure to polychlorinated dibenzodioxins is the skin condition called chloracne. This is an extremely disfiguring skin disease. In mild forms of the disease individuals are affected by numerous small blackheads; however, in some cases the disease is characterized by the appearance of large pustules and cysts.

The condition is extremely persistent; workers contracting chloracne following contamination with dioxins still had the disease 15 years after exposure had ceased.[1] Its rapid onset leads one of the world's leading experts on chloracne, Dr. James Taylor of the Department of Dermatology at the Cleveland Clinic Foundation in Ohio, to comment that its symptoms represent one of the most sensitive indicators of chronic exposure to dioxins.[2] Taylor's view is based on the analysis of medical reports on workers exposed to these chemicals in industry. The first person to actually associate chloracne with dioxin poisoning was a German dermatologist at the University of Hamburg. Dr. Karl Schultz's involvement with the chemical began as a result of an explosion which occurred in 1953 at the trichlorophenol plant owned by the German chemical company Boehringer. Three years after the accident, an employee of Boehringer with a severe form of chloracne was sent to Schultz for treatment. In the course of routine enquiries, Schultz succeeded in identifying a further 30 cases with the skin condition.[3]

Chloracne was first observed by the German doctor S. Von Bettman in 1897.[4] But it was another two years before the possible cause of the disease was suggested, by another German clinician Dr. Karl Herxheimer, who considered that chlorine gas might be the agent responsible for chloracne.[5] Herxheimer proposed, therefore, that the condition be called "chloracne," a shortened form of chlorine acne. In the ensuing 50 years it was realized that chlorine was not the real cause of the problem, but instead the skin condition was caused by exposure to a variety of chlorine- and bromine-substituted aromatic compounds. Chlorine and bromine belong to the group of chemical compounds called halogens. The term *halogens* is derived from the Greek words for sea salt (*hals*) and produce (*gennao*). Three of the halogens—chlorine, bromine, and iodine—are found in the sea as salts, resembling sea salt, which is sodium chloride.[6] Halogens, particularly chlorine and bromine, combine readily with a variety of organic compounds containing a benzene ring. Many of these halogenated compounds cause chloracne. The halogenated chemicals which have this effect include naphthalenes, biphenyls, dibenzofurans, dibenzodioxins, aniline, and azo- and azoxybenzenes.[2,7] Some of these chemicals are described in Chapter 1 on the Chemistry and Occurrence of Dioxins. For the purpose of this narrative, however, discussion of the subject will be brief; readers interested in more detailed information should refer to the four excellent reviews on the subject.[1,2,7,8]

The clinical features of chloracne consist of straw-colored cysts, blackheads, pustules, and abscesses which eventually result in scarring.[2] Chloracne is a disorder of the pilosebaceous mechanism, with the overproduction of keratin in the sebaceous ducts in the skin. This results in the development of the comedome or blackhead seen in all types of acne. Lipid metabolism in the skin of chloracne cases is probably also disturbed. Blackheads, when they appear, give the skin a dirty grey appearance.[9] According to Dr. Kenneth Crow, a world authority on chloracne, mild cases of the disease resemble adolescent acne with the one curious exception, blackheads and cysts tend to concentrate round the eyes and on the ears.[8] Crow says that there is evidence to suggest that different areas of the human skin vary in their response to a standard concentration of chloracnegenic (chloracnecausing) chemicals. The face appears to be the first affected in the manner as already described, that is, the skin around the eyes and on the ears being the most sensitive. In fact those areas are frequently affected when every other part of the skin is normal. In more severe conditions, the disease is apparent in other parts of the body in the following order: the remaining facial area, the neck, shoulders, genitals, chest, and lower trunk. Only in the worst cases, Crow claims, are the hands, feet and legs affected.[8]

Several severe cases were documented in 1949 by two clinicians who studied some of the earliest cases of dioxin-associated chloracne. William Ashe and Raymond Suskind wrote of some of the workers exposed to dioxin at the Monsanto Chemical Company in 1949: "When these men are in a closed room together, there is a strong odour which suggests a phenolic compound, but cannot be identified with certainty. This odour is definitely not the odour of sweat nor the odour of rancid fat in their skin lesions. While we have been unable to prove it, we believe these men are excreting a foreign chemical through their skins."[10]

In a subsequent report,[11] the clinicians described the symptoms of a 36-year-old male worker employed by Monsanto. From his work record it seemed that he had been exposed repeatedly to high concentrations of dioxins, which resulted in the appearance of cysts and blackheads all over his body and face. Shortly after the skin lesions appeared, his skin darkened and remained a grayish-brown color for over a year before it began showing signs of lightening. Ashe and Suskind wrote that the man was handsome and proud of the fact. But he also had strong racial prejudices; his darkened skin, say the authors, meant that on occasions he was "mistaken for a negro which he deeply resents and as a result had secluded himself from most social activity." Ironically on the occasions he was regarded as a black American, on buses and in theaters, he was forced to conform with the racial segregation customs of the area. Not surprisingly, perhaps, Ashe and Suskind wrote that their patient had a "disturbing emotional problem" which required treatment. These cases are discussed more fully in the section on Monsanto in Chapter 6.

Karl Schultz's patients, however, were not affected as severely as this. But their condition was bad enough, and the cause had to be determined. To help discover it, Schultz used a test devised 15 years earlier and which involved the application of chemicals to the ear of a rabbit, the skin of which reacts similarly to human skin with chloracnegenic chemicals.[12] Schultz tested pure 2,4,5-trichlorophenol without effect. However, a commercial grade preparation caused generalized reddening and inflammation to the skin within days, followed a week and a half later by chloracnelike weals. The conclusion was obvious. Commercial grade trichlorophenol had a contaminant which was causing chloracne in the Boehringer workers.

In the course of his work, Schultz collaborated with George Sorge, a chemist in charge of the local chemical laboratory at Boehringer's Hamburg site. Sorge supplied Schultz with a number of compounds—potential contaminants of the trichlorophenol—for testing. Some were chloracnegenic but none could be isolated from the commercial grade phenol.[3]

For a while the clinician and chemist were at a loss to explain the chloracne in the workers. The admission of a research assistant with chloracne to Schultz's department of dermatology sometime later soon reopened the investigation. The assistant had been employed to help the professor in the department of wood chemistry with an investigation of the potential of chlorophenols as wood preservatives. In the course of the work some 20 g of tetrachlorodibenzo-*p*-dioxin (dioxin) had been synthesized and stored in a dessicator. Opening a dessicator can be a hazardous operation as powders are liable to "puff" when the seal is broken. Just such an accident occurred to the assistant: a cloud of dioxin wafted into his face when the dessicator was opened.

This unfortunate incident provided the clue Sorge was after. Dioxin at this time was an unknown entity. Sorge, however, recognized that it could be the contaminant he and Schultz had been looking for. He, too, prepared dioxin in the laboratory, comparing it with the product he had isolated from commercial grade trichlorophenol; they were identical. It seemed as though the cause of chloracne in the Beohringer workers had been found. All that remained was for Schultz to test the dioxin on rabbits. The result was positive. Dioxin was highly active in the test. As final confirmation that dioxin was the contaminant they were seeking, Schultz applied the chemical to a spot on his lower left underarm. In two weeks chloracne had developed and Schultz and Sorge knew they had finally found the cause of the skin disease in the Boehringer workers.

Even before details were published of the formation of dioxin during the heating of chlorophenols, Sorge realized that the amount of contaminant produced in any reaction to make 2,4,5-trichlorophenol depended on temperature.[13] The higher the temperature the more dioxin is produced. Sorge, therefore, modified Boehringer's manufacturing process to one using a low-temperature procedure and as a result no new cases of chloracne have been reported.

The lessons Boehringer learned are instructive to all. Realizing that its old, 2,4,5-trichlorophenol reactor was contaminated with dioxin, the company dismantled the plant and disposed of its contents. Boehringer, according to Professor Bo Holmsted[3] of the Karolinska Institute in Stockholm, also informed other manufacturers of trichlorophenol of the risks involved in their production of the chemical and how they could be avoided. It is obvious, however, that Boehringer's experience did not lead to a containment of the problems associated with the chemical process; there were many more accidents still to come.

But Schultz and Sorge together had made a great advance for occupational hygienists even though few availed themselves of the information. Schultz eventually gained the credit for identifying dioxin as the cause of chloracne and published his findings.[14] Sorge, restricted by Boehringer company policy, could not openly acknowledge his role in the investigation.[3] His consolation must have been that of the knowledge of a job well done. It was now clear that chloracne was an early warning sign of a trichlorophenol reaction gone wrong. The writing was on the wall.

References

1. Taylor, J. S. A continuing problem, *Cutis* **13**, 585-591 (1974).
2. Taylor, J. S. Environmental chloracne. Update and overview, *Annals of New York Academy of Sciences* **320**, 295-307 (1979).
3. Holmstedt, B. Prolegomena to Seveso, *Archives of Toxicology* **44**, 211-230 (1980).
4. Bettmann, S. Chlorakne, eine besondere form van professionaller hauterkrankung, *Deutsche Medizinische Wochenschrift* **27**, 437 (1901).
5. Herxheimer, K. Weitere mitteilungen uber chlorackne, in *VII Dermatological Kongress*, Breslau, p. 152 (1901).

6. *The Halogens in Mellors Modern Inorganic Chemistry.* Revised edition by G. D. Parkes and J. W. Mellor, Longmans, Green & Co., London (1943), p. 483.
7. Crow, K. D. Chloracne, a critical review including a comparison of two series of acne from chloronapthalene and pitch fumes, *Transactions of St. Johns Hospital Dermatological Society* **56,** 79-99 (1970).
8. Crow, K. D., Chloracne: the chemical disease, *New Scientist* **1098,** 78-80 (1980).
9. Schultz, K. H. Clinical picture and etiology of chloracne, *Arbeits Med. Socialmed. Arbeitshyg* **3,** 25 (1968).
10. Ashe, W. A., and Suskind, R. K. Report—Patients from Monsanto, Chemical Company Nitro, West Virginia (5 December 1949).
11. Ashe, W. A., and Suskind, R. K. Progress Report—Patients from Monsanto, Chemical Company Nitro, West Virginia (April 1950).
12. Adams, E. M., Irish, D. S., Spencer, H. C., and Rowe, V. K. The response of rabbit skin to compounds reported to have caused acne from dermatitis, *Industrial Medicine Industrial Hygiene Section* **2,** 1-4 (1941).
13. Sandermann, W., Stockmann, H., and Cartn, R. Pyrolysis of pentachlorophenol, *Chemische Berichte* **90,** 690-692 (1957).
14. Kimmig, J., and Schultz, K. H., Chlorierte aromatische zyklische. Ather also Ursache der Sogenannten Chlorakne, *Naturwissenschaften* **44,** 337-338 (1957).

6
Industrial Accidents in Trichlorophenol Manufacture

The complications involved in the assessment of the risk to humans following exposure to polychlorinated dibenzodioxins (PCDDs) are numerous. The principal difficulty is the absence of any hard data on the actual concentration of PCDDs to which individuals were exposed. In some cases exposure to the dioxins may have been an isolated event, but in the majority of instances contact with the chemical probably occurred over a prolonged period. In addition to which individuals would probably have been exposed to a number of other chemicals at the same time. Introducing other factors into the risk equation has the effect of "confounding" the overall picture—making interpretation of any result far more involved.

Many believe that the clue(s) to the question "Are dioxins dangerous to humans?" lies in the studies being made of the health problems of workers exposed to these chemicals in industry. The number of workers reported to be in this category is about 2000; some are now deceased. All have or had been involved in the manufacture of 2,4,5-trichlorophenol at one time or another since 1949. The group represents, therefore, the largest known human population exposed to these chemicals. Exposure in all cases occurred either as a result of explosions in the reactor vessels housing the trichlorophenol mix, or through poor hygiene in the factory (occupational exposure). A list of the known industrial accidents is shown in Table XI.

In later sections some of the more important industrial accidents and their consequences are discussed in more detail.

The accident at Seveso is also included in the league table of accidents. It, too, is discussed at length elsewhere in this book. At this juncture it is only necessary to point out that no industrial workers employed at the ICMESA works near Seveso were affected by the explosion. The victims in this case were the local population.

TABLE XI. Accidents in Chemical Plants Involving the Manufacture of Chlorinated Phenols[a]

Date	Country	Manufacturer/location	Product	Cause of exposure	Personnel affected	References
1949	United States	Monsanto/Nitro, West Virginia	TCP	Explosion	228	4–7
1949	Federal Republic of Germany	Nordrhein-Westfalen	PCP, TCP	Occupational	17	8
1952	Federal Republic of Germany	Nordrhein-Westfalen	TCP	Occupational	60	9
1952–1953	Federal Republic of Germany	Boehringer	TCP	Occupational	37	3, 10
1953	Federal Republic of Germany	BASF/Ludwigshafen	TCP	Explosion	75	11–14
1953–1971	France	Rhone-Poulenc/Grenoble	TCP	Occupational and explosion	17	15
1954	Federal Republic of Germany	Boehringer, Ingelheim/Hamburg	TCP, 2,4,5-T	Occupational	31	16–18
1956	United States	Diamond Alkalai/Newark, New Jersey	2,4-D, 2,4,5-T	Occupational	29	19, 20
1956	United States	Hooker/Niagara Falls, New York	TCP	Occupational	Many	1
1959	Italy	Industrie Chimiche Melegnanesi Saronio/Milan	TCP	Occupational	5	21
1959	United States	Thompson-Hayward/Kansas City, Kansas	TCP	Occupational	—	1

Year	Country	Company/Location	Chemical[b]	Type	Number affected	Ref.
1960	United States	Diamond Shamrock	TCP	Occupational	Many—1 fatal	1
1963	Netherlands	Philips-Duphar/Amsterdam	TCP	Explosion	106	14, 22–24
1964	USSR		2,4,5-T	Occupational	128	25
1964	United States	Dow Chemical/Midland, Michigan	2,4,5-T	Occupational	30	26
1964–1969	Czechoslovakia	Spolana	TCP	Occupational	80, 2 fatal	27–29
1968	United Kingdom	Coalite and Chemicals Products/Bolsover, Derbyshire	TCP	Explosion	90	30
1970	United Kingdom	Coalite and Chemicals Products/Hertfordshire	TCP	Occupational	3	31
1970	Japan		2,4,5-T	Occupational	25	32
1972	USSR		TCP	Occupational	1	33
1972–1973	Austria	Linz Nitrogen Works	2,4,5-T	Occupational	50	1
1974	Fed. Rep. of Germany	Bayer/Uerdingen	2,4,5-T	Occupational	5	1
1976	Italy	ICMESA/Meda/Seveso	TCP	Explosion		34–36
1976	United Kingdom	Monsanto/South Wales	PCP	Occupational		37

[a] This information is compiled using information from three sources (References 1–3). Original references, where available, are quoted.
[b] TCP, 2,4,5-trichlorophenol; PCP, pentachlorophenol; 2,4-D, 2,4-dichlorophenoxyacetic acid; 2,4,5-T, 2,4,5-trichlorophenoxyacetic acid.

Monsanto, U.S.A., 1949

With dioxin known to cause cancer in animals it must be assumed that it could cause cancer in humans. Is there any evidence that it does? Such evidence, as it exists, is far from clear. Investigations involving workers exposed to dioxin in industry present a confusing picture at present. One study suggests that there is a cancer risk from dixoin,[38] another suggests, cautiously, that there might not be.[39]

The latter study was carried out on workers exposed to dioxin in 1949 at the West Virginia site of the Monsanto Chemical Company; 122 workers were exposed to the chemical as a result of an industrial accident involving a runaway reaction in the 2,4,5-trichlorophenol plant. An additional 150 workers were exposed to, and affected by, dioxin in the routine handling of the herbicide 2,4,5-T. This second group was contaminated between 1949 and 1968.[40] In the early days of trichlorophenol production the process was said by one scientist to have been extremely crude with workers removing chemicals from the reactor by scraping it out with their hands.

Latency periods between exposure to a chemical carcinogen and the development of a tumor vary, but in general are of the order of 20–30 years. To assess the cancer risk to humans of exposure to dioxin requires investigations of incidents where the chemical affected individuals at least 20 years ago. The Monsanto accident occurred over 30 years ago. The clues to be gathered from the medical histories of those exposed to dioxin at this company should answer the question "How harmful is dioxin?"

"Very" might be the reply a worker exposed to the chemical would give. Dioxin's cancer threat to humans may be unclear, but its effects on the skin are clearly visible. Chloracne, the disfiguring skin condition which dioxin causes, was particularly severe in the Monsanto workers. The problems which one worker experienced in this regard are detailed in Chapter 5. The difficulties some of the workers faced are worth recounting, as are the circumstances surrounding the accident.

On 8 March 1949 the safety valve and pipe connections on one of three autoclaves in Building 41 on the Monsanto site were blown open releasing a fine black powder and a thick, sticky, dark brown substance which covered the interior of the building.[41] The cause of the rupture was an increase of pressure in the autoclave. A runaway reaction in the 2,4,5-trichlorophenol reactor had led to a rapid increase in temperature. Pressure in the vessel built up rapidly and eventually blew the safety valve and ruptured pipe connections.

Within an hour, workers were back in the building to assess the damage and clean up. Few safety precautions were taken by those entering the building; it was just another minor accident. One of the first to return to

Building 41 was a 32-year-old steamfitter, Ivan McClanahan. For the three hours he was in the building, McClanahan experienced a burning sensation of the eyes, nose, and throat. He was sent back in the following day before being transferred to another part of the plant.

Several hours after his first exposure McClanahan developed a severe headache. He had also experienced a feeling of nausea. Later that evening, he had difficulty in standing up and suffered from vertigo. The nausea felt earlier became more serious as McClanahan began to vomit. This bout passed but the headache persisted; it lasted eight days and recurred at intervals thereafter.

McClanahan's skin symptoms occurred six days after the accident. His face became inflamed and there was a marked swelling over the eyelids, nose, and lips. A few days later pustules, comedomes (blackheads), and cysts began to appear on his face, forearms, shoulders, neck, and trunk. His skin condition continued to deteriorate with the blackheads and cysts increasing in number. McClanahan's skin became pigmented and he complained of being easily fatigued and feeling weak.

A month after his exposure to the dioxin-contaminated material in Building 41, McClanahan experienced aches in his thighs and calves. Exertion only aggravated the condition and he was virtually immobilized for a number of weeks. In all, the pain lasted for about six weeks, but continued to recur on occasions. When he was examined some months after the accident, McClanahan said he still felt weak. He found mild to moderate exertion was fatiguing, his sleep was still disturbed, and, perhaps not surprisingly, he felt nervous.[41]

Thirty-six-year-old Paul Willard, the chief operator in Building 41, had similar symptoms. Six weeks after resuming his post in the accident-damaged building Willard noted a rash on his right leg; this was followed by an aching pain. A few days later his left leg was similarly affected. Blackheads, cysts, and pustules began to appear on his face and neck. At first the lesions were mild but as the weeks passed the condition became much more severe.

Willard found it impossible to work in the contaminated building and not be covered in the explosion residue. The continuous exposure had severe consequences for his health. The pain in his legs became more intense and in early July 1949 he was hospitalized; he was unable to walk. He remained there for 16 days until the pain subsided. A few months later he was readmitted to a hospital. Willard was now experiencing severe pain in the chest, back, and neck; his breathing was labored. With treatment the pain subsided and Willard returned home. He, too, when examined some eight months after the accident, complained of insomnia and feeling unduly nervous.[41]

Two other patients, Jesse Steele and 56-year-old Jonathan Hurley, examined at the same time had almost identical problems.[41] Chloracne was common to all four. So too was an odor; according to two doctors[41] who examined the men, when they were in a closed room together there was a strange odor familiar to that of a phenolic compound. The smell was not that of sweat nor the odor of rancid fat in their skin lesions. The doctors believed that the workers were excreting a foreign chemical through their skins.

There were other problems too. Peripheral neuropathy (nervous tissue damage) was detected in one of the four following examination of a tissue biopsy sample. All four men had raised blood fat levels, total blood cholesterol being particularly high in two cases. These changes in blood fat levels are not unusual; high blood cholesterol levels have been observed in other workers exposed to dioxin.[42,43] Four months later nine workers, including the original four, were examined by Drs. William Ashe and Raymond Suskind. The picture had altered little. Although treatment for the chloracne (twice daily washes with neutral soap and water, application of lotio alba [skin lotions], daily hour washes, 200,000 IU of Vitamn A daily, a general high protein, high vitamin diet, and draining of cysts and pustules) had improved the skin condition in the four men seen earlier, the treatment had little impact on blood lipids. Fasting blood samples taken from the nine showed eight to have increases in total blood fat. Four out of seven subjects had increases in total blood cholesterol.[44]

In their second, slightly larger study Ashe and Suskind summarized some of the additional findings on the nine men. These included respiratory difficulties, limited to shortness of breath and wheezing; neurological changes, characterized by localized pain and weakness; central nervous system disturbances such as irritability, nervousness, and insomnia; and finally a loss of libido together with some impotence.

Although Ashe and Suskind were able to record a marked improvement in their patients' clinical condition, it is known that the skin condition was difficult to treat. A study,[45] conducted in 1979, of the prevalence and severity of the chloracne in the dioxin-exposed Monsanto workers revealed that many still had mild to moderate forms of the disease—the disfiguring skin condition had persisted for over 30 years.

Although chloracne is at times persistent, improvements in the condition of the skin can be monitored. Assessing other factors like increased risk from heart attacks due to high blood cholesterol levels or even the likelihood of contracting cancer is far more difficult. Both of these risks are real for workers exposed to dioxin and particularly so for the Monsanto men. The high blood cholesterol values in this group of workers put them at greater risk of developing cardiovascular problems. The cancer threat was there too. Dioxin causes cancer in animals and could do the same in humans. For many chemical carcinogens, the time between exposure to the chemical and the appearance of the tumor is often 20–30 years. Thus if dioxin does cause cancer in humans, the Monsanto group is a vital population to study.

Such an investigation has been conducted, and the results are more reassuring than might have been expected—121 workers exposed to dioxin in the accident in Building 41 have been traced; 32 men have died since the accident. The number who might have been expected to die based on national figures was 46.41. Thus the mortality ratio (standardized) for all deaths is 0.69. Breaking down the figures, the mortality ratio for deaths due to cancer is 1; 9 deaths were recorded in the group, 9.04 were expected. The mortality ratio for deaths due to circulatory diseases was low at 0.68; 17 deaths occurred in the Monsanto men; 25.01 were expected.[39]

The number of deaths from cancer in this group of workers is no higher than the norm, as the authors point out. However, there may be more to note in these cases than the figures suggest. Of the nine deaths recorded, one was reported to be due to a soft tissue sarcoma (a malignant fibrous histocytoma). Another death from a soft tissue cancer in a different group of Monsanto workers exposed to 2,4,5-T has also been reported; in this group of men one died from what is termed "a generalized liposarcoma."[46] Two workers exposed to dioxin at the Dow Chemical Company have also contracted soft tissue cancer,[47] one of the two men has died.

For the majority of soft tissue cancers the cause of the disease is unknown. The tumors are rare, hence the interest which has been taken in their incidence among dioxin exposed workers. In 1975 the incidence of soft tissue sarcomas in U.S. males aged 20–84 years was 0.07%. Taking the Monsanto and Dow workers as a whole, 3/105 deaths were due to these sarcomas.[47] This represents an incidence of 2.9% and some 41 times greater than the apparent norm. It is perhaps worth noting that all four of the men who developed soft tissue cancer smoked. It has been suggested, therefore, that the risk of developing cancer may be greater in those dioxin exposed workers who smoke.[47]

The authors of the Monsanto mortality study—Judith Zack, an epidemiologist employed by Monsanto, and Dr. Raymond Suskind, now director of the Institute of Environmental Health at the University of Cincinnati Medical School—point out, however, that the number of both workers studied and deaths recorded in their study are small and that the study "cannot be considered conclusive." Epidemiology studies require studies on large numbers of individuals. Definitive information on the risk dioxin-exposed workers run will thus require the pooling of information from all the accidents where the chemical was released.

Monsanto itself will provide further numbers for this assessment. In addition to the 121 workers exposed to dioxin in the accident, more than 150 were affected by the chemical between 1949 and 1968 in the routine manufacture of 2,4,5-T. These men, according to Suskind, were also being traced.[40] Zack and Suskind have also conducted a large mortality study on 884 men employed at Monsanto at its Nitro, West Virginia, plant.[48] The study group represents the Nitro plant's total workforce. Of the men selected for study, 721 were still alive, and of the 163 deceased, 58 had been exposed to 2,4,5-T and potentially to dioxin. The study indicated that the number of deaths due to cancer in the group was lower than would be expected. However, the number of deaths due to circulatory disorders (largely those involving the heart) was slightly elevated. Actuarial figures for mortality rates in the total U.S. male population suggest that 26.48 circulatory-related deaths would be expected in the period under study, whereas the actual number of deaths recorded was 31. The authors point out, however, that the difference between the expected and the observed heart attack deaths is not statistically significant, and probably reflects the higher than average mortality rate for heart disease in the country in which the plant is sited.

West Germany, BASF, 1953

On 17 November 1953, one of the first European trichlorophenol accidents occurred.[51] The reactor which exploded belonged to the giant West German corporation Badische Anilin & Soda-Fabrik (BASF) and was located at the company's main site in Ludwigshaven, on the banks of the Rhine. All chemical companies require an adequate supply of water for their reaction processes, for washing and for cooling. The Rhine site was thus ideal for BASF. BASF now has a multinational operation and can be considered one of the world's chemical giants,[49] but parts of its history have not always been so prestigious.[50]

Two alternative solvents are used in trichlorophenol manufacture: methanol and ethylene glycol. The use of methanol has one major drawback; it must be used under fairly high pressures, about 20 atm (atmospheres), whereas ethylene glycol is used at normal atmospheric pressure. BASF used methanol in its trichlorophenol process. On 17 November 1953 an exothermic reaction started in the reaction vessel. Both temperature and pressure in the vessel rose rapidly until the autoclave finally exploded. The contents of the autoclave were ejected with such force that the room holding it was immediately filled with vapor as were three other floors in the building connected to the autoclave room by a single stairwell. Within a few minutes the vapor had cleared; precipitate was clearly visible on the floor, on machinery, walls, windows, and doors.

When workers entered the area to assess and deal with the damage, 42 men were involved in the exercise. All later developed chloracne. Some showed symptoms of severe poisoning. The appearance of chloracne forced the management to provide protective clothing for those workers involved in the refurbishing of the plant. It was later admitted[52,53] that attempts at decontamination had been inadequate. Repeated whitewashing of the walls, spraying of machinery with chloride of lime and water, coating of pipes with rust-proofing agents and iron oxide, spraying of floors with a silicon solution, and the lining of walls with foil and paint failed to prevent the spread of chloracne. Five years after the explosion, one worker in the plant still developed chloracne. The source of contamination was traced back to the autoclave used in the trichlorophenol process. BASF had, in the interim, ceased production of the phenol, repaired and washed the autoclave, and adapted it for use on another process. With the appearance of yet another case of chloracne—75 workers eventually developed the skin complaint—the company had to conclude that its washing of the autoclave with sodium hydroxide, lye soap, and organic solvents had been inadequate to remove all traces of toxic chemicals.

Although today BASF prides itself on the efficiency of its occupational hygiene department, in 1953 the then medical department was not quite so alert. It was not until three weeks after workers first entered the building that the plant doctors became aware that those men working in the contaminated plant were experiencing ill health. As more workers reported to the medical department with chloracne it was decided to introduce rabbits into the contaminated building and to observe any toxic effects. None were seen. The management concluded, therefore, that all was well. Work resumed in the plant but before long a further three severe cases of dioxin poisoning were seen. This led to the initiation of a second rabbit experiment, and animals were housed in the autoclave itself. At first none of the rabbits exhibited any ill effects but five days after their removal from the vessel, two showed definite signs of illness. They were killed and a postmortem examination revealed liver necrosis. At this time it was known that chlorinated naphthalenes could cause necrosis and the management was once again faced with further evidence that their plant remained contaminated.

The presence of the dioxin contaminant in trichlorophenol was not known at the time of the accident, but it is clear that the 42 men who reported ill after cleaning operations in the trichlorophenol plant were in some way affected by the chemical. The only paper available on the clinical symptoms seen in the 42 workers is not as clear as it might be. Few objective clinical measurements are provided for the cases documented. The author refers to

42 workers reporting ill, only half of whom had any visible skin disease (in most other dioxin poisoning accidents the skin diseases preceded the onset of other symptoms). Symptoms seen in the workers were more severe. Some of those with symptoms of poisoning required a long period of convalescence—in the case of one worker, for a year and a half.

Dioxin contamination was not restricted to the workers on the plant as one report makes clear. In a pattern to be repeated following the accident at the Coalite plant in the United Kingdom, some members of the families of workers subsequently developed chloracne. One worker who suffered severe chloracne, toxic polyneuritis (damage to the nervous system) with damage to the motor system, and demonstrable disturbances in his ability to hear, taste, and smell, was the unwitting transmitter of the chemical to his son. The boy, who had slept in his father's bed, used the same bath towels, and wore the father's scarf (following its having been washed three times) developed chloracne on his face. Yet another case of dioxin transmission occurred in the factory canteen. The colleagues of a worker employed in the trichlorophenol plant developed chloracne after the two had sat next to one another during a meal.

In 1969, 18 years after the accident, chloracne was still visible on the face, penis, and scrotum of some of the workers.[52] The first recorded death in the workers exposed to dioxin occurred in 1959. A 57-year-old man, working in the autoclave room five years after the accident, removed his face mask to wipe away sweat caused by the steamy atmosphere. Some days later he is reported to have developed skin lesions, and chloracne was eventually confirmed. Nine months later when the man died the chloracne had not abated. An autopsy report established pancreatitis as the cause of death, the man having widespread necrosis of adipose tissue, a perforated stomach and a duodenal ulcer extending into the pancreas. As animal experiments had established that dioxin affects the enzyme lipase—derived from the pancreas—the coroner felt justified in attributing the cause of death to exposure to the chemical. This verdict was never challenged, although today it is less likely that dioxin would have been implicated so readily on the basis of the evidence provided in 1959.

More recent information about the health of these workers was discussed at an International Agency for Cancer (IARC) meeting at Lyons in January 1978. At that time, 25 years after the accident, BASF appeared extremely worried about the possibility of information on the health effects of the workers being divulged. Dr. R. Frentzel-Beyne of BASF, when presenting his information to the IARC meeting, requested that the data be treated in confidence as the company was concerned about future compensation claims.[54] The company has obviously had second thoughts about their policy as details about the accident have, however, been published.[51,55]

Of the 75 workers who were contaminated with dioxin in the wake of the trichlorophenol reactor explosion, all have been traced.[51,55] The recent investigations on this group have been mortality studies. The 75 have been compared with an internal control group in the factory matched for age and entry into the plant, but selected at random. In January 1978, after 24 years of observation of the exposed group, some 17 workers have died. The comparable figures for Ludwigshaven is 25, and for West Germany as a whole, 16. However, of the deaths 6 were due to cancer, 4 of which involved the gastrointestinal tract. These figures are highly significant for they represent twice the expected rate for any of the control groups. More significant, however, is the fact that 2 of the gastrointestinal cancers occurred in workers between the ages of 65 and 69. Compared with age-matched controls the recorded deaths are 10 times higher than expected. A second epidemiological study of this group of men conducted 3 years later comes to the same conclusion.[56] Commenting on the incidence of stomach cancers in the dioxin-exposed workers, the authors say that it was "... considerably greater than expected and cannot be adequately explained as a mere chance event."

There is concern that dioxin may cause genetic damage and that it may damage chromosomes. BASF workers exposed to dioxin have been examined for any increase in the incidence of chromosome abnormalities. The figures suggest that none has occurred. In an as yet unpublished study, Professor Gunter Rohrborn of Dusseldorf University's Institute for Human Genetics and Anthropology reports a 4.4% incidence of chromosome breaks and gaps in the dioxin-exposed workers, compared with a figure of 2.25% for a control population. The differences between the two populations have been evaluated statistically, and, according to Rohrborn, are not significant.[57]

The BASF figures for the two epidemiological surveys were the first to reveal that workers exposed to dioxin may be at risk of developing cancer. In view of the confirmatory animal studies which show dioxin to be carcinogenic, the BASF figures now confirm the need for caution and a careful monitoring of all dioxin exposed groups.

Only one accident involving a trichlorophenol reactor is documented prior to 1953, that at Monsanto in 1949. Since then there have been many more and the number of workers known or presumed to have been exposed to dioxin is nearly 2000. It would seem to be in their interest that figures on mortality in dioxin-exposed groups be collected. The IARC has agreed to do the collation, but a year after initiating the program the Agency had little to show for its efforts. No information had been communicated to it by the various industries attending the IARC meeting. With no new information available to it the IARC was unable to be of much assistance to Italian scientists and clinicians concerned with the health of the Seveso population—one of the initial objectives of the meeting in Lyons in 1978.[58]

Philips-Duphar, Holland, 1963

On 6 March 1963 Philips-Duphar, an Amsterdam-based firm engaged in the manufacture of trichlorophenol, fell foul of the exothermic reaction which had caused explosions in so many other trichlorophenol reactors. In this case the reactor went out of control at the start of the operation, rather than at the end, which had been the pattern of previous accidents. The rapid escalation in temperature and resulting pressure increase blew the safety valve, ripping the lid off the autoclave. When the blast occurred, the temperature was estimated to have been about 400–450°C and the pressure greater than 80 atmospheres. The report of the investigation into the accident concluded that mechanical failure was not to blame, and said that operators' error could not be ruled out.

Aware of the consequences of previous accidents in trichlorophenol plants in the United States and West Germany, the Philips-Duphar management adopted the only prudent course; they closed the building, intending to carry out thorough decontamination. Although the management was aware that something particularly noxious was produced during the reaction process they were unaware—as were some other manufacturers—of the presence of dioxin. This revelation did not come until three years later when the Dow Chemical Company first isolated and then developed a method for measuring the chemical using gas liquid chromatography.

When Duphar was assessing the extent of the contamination of its premises, it used what it considered to be the best analytical techniques available. This involved the use of a spectrophotometer. In the presence of light of a particular wavelength all compounds exhibit a characteristic absorption spectrum which, like a fingerprint, can be used for the purposes of identification. However, the spectrophotometric peak which Duphar so assiduously monitored to assess the effectiveness of its decontamination process was, during an investigation conducted some years later, identified not as dioxin but as a phenolic compound—by no means as toxic or as persistent as the dioxin contaminant.

As well as employing the spectrophotometric method for monitoring decontamination Philips-Duphar also conducted animal experiments for the same purpose. Rats were housed in cages inside the building; others had soot taken from the autoclave following the explosion, incorporated into the feed, or applied to the skin. Upon examination 11 days later, animals in all three groups showed lipid accumulation in the liver—a sign of poisoning. A second skin experiment showed similar results. At the end of March, nearly four weeks after the explosion, rats were again placed in the building to expose them to chemicals by inhalation. When examined three weeks later none of these animals had the lipid accumulation noted previously.

However, a casualty in a group of six rabbits placed in the building at the same time showed that all was not well. One week later, one of the six animals died. The management had to conclude that the building was still contaminated. Three months later yet another rat experiment was carried out. Animals kept in open cages were deliberately exposed to paint being removed from machines and the walls of the building. Of the 48 animals present at the outset six died. An autopsy revealed atrophy of the thymus gland, enlargement of the testes, and the familiar accumulation of lipid in the liver.

In the wake of Dow Chemical's announcement of a method for measuring dioxin Duphar used the procedure for checking its own plant. The investigation showed dioxin to have impregnated the walls in amounts of 1000 ppm (parts per million). Using these figures the company estimated that a considerable amount of dioxin, 200–250 g (grams), was released in the explosion. There was a further shock in store for the management. The cleaning process involving the use of water, a surface-active compound, a weak solution of sodium hydroxide, and thorough ventilation was soon found to have been inadequate; an estimated 20–200 g of dioxin was still present in the building. It was hoped that reconstruction work together with cleaning would be sufficient to decontaminate the plant but at the end of 1973 when reconstruction work had been completed and damaged machinery replaced, a final animal toxicity study still confirmed the presence of dioxin. For the management this was the end of the road, and the decision was taken to close, and subsequently to demolish, the building.

Demolition was carried out under the supervision of Dr. L. M. Dalderup, the medical adviser to the Factory Inspectorate in the Amsterdam district. Dalderup's prime concern was to keep dust to a minimum and to prevent demolition workers from being exposed to toxic material. All who were employed on the site were required to wear protective clothing consisting of an air-supplied PVC (polyvinyl chloride) or neoprene suit with integral helmet and gas-tight zipper together with well-fitting boots. Entry to and from the site was via a specially constructed building in which workers both kitted themselves out for work on the plant, and decontaminated and washed themselves on leaving it. This auxiliary building for access to the site was later adopted as the model to be used in the decontamination work at Seveso.[59]

For advice on the procedures to use in the demolition of the buildings Dalderup turned to BASF in Ludwigshaven; this company's reactor accident had occurred ten years prior to the Philips-Duphar explosion. BASF responded to the request by providing details of its own decontamination procedures. Dalderup faced a particular problem caused by the actual site of the Philips-Duphar building which was invariably swept by prevailing westerly winds. To keep dust to a minimum, all masonry was thoroughly wetted. The walls were demolished using an 800-kg (kilogram) iron ball, swung so as to loosen the brickwork without allowing it to fall. Chunks were mechanically grabbed and deposited directly into the hold of one of three barges (each weighing about 330 tons) requisitioned for the disposal of the contaminated material. A new harbor was required for the vessels so that when berthed they would lie parallel to the longest walls and separated from the nearest by only the width of a crane track. When filled the barge holds were hermetically sealed. The vessels were finally towed out to sea with their toxic cargo.[60] At a suitable deep section of the Atlantic the barges were sunk.[61,62] Such drastic methods removed the source of contamination but did not signal an end to Philips-Duphar's involvement with the consequences of the accident. Company and outside contract workers had been exposed to dioxin in the wake of the explosion. One hundred and six men had been employed in the clean-up process between March and July 1963. Of these, 44 were factory employees—exposed during every phase of the cleaning and reconstruction work—18 belonged to an outside contract cleaning firm, and the remaining were nonfactory employees engaged for plumbing, painting, and insulation work. Workers in the last two groups were exposed to dioxin between May and July 1963.

Two workers employed in the building at the time of the explosion developed minor skin lesions on the following day and 4–6 weeks later had the characteristic signs of the disfiguring skin disease chloracne. The 44 factory employees engaged in decontamination work were all medically examined at intervals until March 1964 and in a number of cases for a year after this; 26 of them developed chloracne. By 1977 only 27 of these workers were in the factory's employ. On reexamination, 20 still had unmistakable signs of the skin disease. Of the other 17, 3 have since died. One, a 69-year-old former employee, died of a heart attack in 1977, and a 76-year-old was killed in a road accident in the same year. A 54-year-old worker died from carcinoma of the pancreas in 1964; he had also suffered from chloracne.

Only 16 of the 18 cleaners from the contracting firm were medically examined in 1963. Of this group 10 had developed chloracne. Fourteen years later in 1977 the only additional medical evidence concerning this group was that 4 had died. The only data available on any of them shows that at least 3 of the deaths were due to myocardial infarcts (heart attacks).

In the last group of 44 nonfactory workers 16 were examined in 1963; 8 had chloracne. In 1977, 32 from this group were traced when one death, from a myocardial infarct, was recorded. The total deaths from "natural causes" by 1977 thus stood at 7. Mortality data from the Amsterdam district indicate that among workers 7 deaths in a population of 93 in the age range concerned is not excessive. Unfortunately there is no matched control group for this population so any conclusions are, at best, speculative. It has been noted,[63] however, that if the group of cleaners is considered separately then the 4 deaths recorded for workers aged 41-65 is above the statistical norm. The high incidence of chloracne in the cleaners was also unexpected as these workers entered the dioxin-contaminated building at a later date than the regular employees. The use of organic solvents to remove paint may have, inadvertently, exposed the cleaners to more dioxin than usual,[63] a factor which could account for the high incidence of skin disease. Mortality statistics for the cleaners are further complicated by the fact that, by virtue of their profession, these workers were exposed to a variety of toxic substances both before and after coming into contact with dioxin.

For the future, it has been proposed that the health of the 27 employees still at Philips-Duphar be monitored and the data on the group compared with that from a matching control population. Data on mortality is also to be kept for all those workers exposed to dioxin and traced in 1977.[63]

There is an ironic postscript to the Philips-Duphar incident. In 1980 it was revealed that the care which had been taken over the disposal of the dioxin-contaminated factory had not been observed when disposing of other waste products similarly polluted. When it had been making 2,4,5-trichlorophenol on a regular basis the company had routinely disposed of dioxin-contaminated residues—present after every production run—in the nearest waste tip. In the case of Philips-Duphar, the tip was sited at Volgermeerpolder, some 5 miles from the center of Amsterdam.[64]

With some of the metal drums having been on the site for 20 years or more, corrosion was inevitable. The corrosion was followed by leakages, and as proof of this dioxin has been detected in the surrounding topsoil. The concentration of the chemical in the soil is estimated to be from 0.44 to 7.0 μg/kg dry weight of soil.[65] The higher figure is similar to that encountered in the dioxin-contaminated areas of Seveso from which residents were evacuated.

Coalite and Chemical Products Limited, U.K., 1968

In the early hours of 24 April 1968 the trichlorophenol reactor of the Fine Chemicals Unit of Coalite and Chemical Products Limited exploded, killing the duty chemist on the night shift. The 38-year-old chemist, Eric Burrows, was one of eight workers in the building at the time of the explosion. Burrows was the only worker killed in the accident; the other seven, suffering from shock, were hospitalized.[66]

The major part of Coalite's operation is concerned with the production of solid smokeless fuels for domestic uses. Coalite is no chemical giant, but then again, it is not a small company. In its annual report for 1977/1978, Coalite states that it employs over 7500 people and has a turnover in excess of £168.3 million.[67]

Manufacture of trichlorophenol at Coalite began in August 1965 at the Company's Bolsover Fine Chemicals Unit. As with most trichlorophenol producers worldwide, Coalite used the chemical as the starting point for the production of the herbicide 2,4,5-T. Givaudan and its parent company Roche, which owned the trichlorophenol plant at Seveso, were the only group which used the phenol to make the bactericide hexachlorophene.

The explosion at Coalite has been well documented in the scientific literature. Both the theory[68] as to the cause of the accident and the consequences in terms of the health of the plant workers[69] have been reported. Trichlorophenol at Coalite was produced in an autoclave by reacting 1,2,4,5-tetrachlorobenzene with caustic soda in the presence of two solvents, ethylene glycol and orthochlorobenzene. The latter solvent was used simply to prevent blocking of the condenser by solid tetrachlorobenzene. The autoclave was heated by oil, and the reaction to produce trichlorophenol occurred at 180°C. Sampling of the autoclave contents was used to monitor the reaction's progress. On completion, and before transferral, the vessel's contents were cooled to 140°C. However, at midnight on 23 April 1968 this sequence of events went awry.

In a preamble to a report[69] on the health of the Coalite workers, the company general practitioner, Dr. George May, wrote that "... at midnight on 23 April 1968, the reaction temperature reached 175°C and thereafter it rose continuously for 50 minutes. When it reached a point somewhere in excess of 250°C an explosion of considerable violence occurred and the supervising chemist was killed by falling masonry." Subsequent research[68] has revealed the explosion to have been a two-stage process. At 225°C the reaction became exothermic—generated its own heat—which in the closed reactor vessel led to a buildup in pressure. This buildup led to the eventual rupture of the vessel and to the release of ethylene glycol and orthochlorobenzene vapors into the atmosphere. In the presence of atmospheric oxygen, these vapors form an explosive mixture requiring only an ignition source to detonate them. In the opinion of the investigation into the cause of the accident from the Ministry of Technology and the Factories Inspectorate, an overhead electric light is thought to have been that source.

The inquest on the duty chemist was held at Chesterfield on 16 May 1968. In his evidence, the coroner, Mr. Michael Swanick, pronounced himself satisfied that there had been no breach of regulations on the part of any one individual.[70] Recording a verdict of accidental death, the jury recommended an audible warning signal—absent on the reactor which exploded—to indicate excessive pressure in reactors so that swifter remedial action could be taken to prevent an explosion occurring.

The whole unit was closed by the management in the wake of the accident and investigations into its cause begun. At the same time the employees present in the building at the time of the accident were given medical examinations. Initial clinical findings on these men suggested some abnormal white cell counts, suggesting a defective immune system, and in three cases, an elevation of glucose in the urine, suggesting possible liver and kidney damage. A repeat screening ten days later did not confirm this pattern, however, for all the tests gave values within normal limits.

Coalite's management thought the results reassuring and reopened the unit keeping the trichlorophenol section sealed off for the decontamination at a later date. According to May, all workers were closely monitored for any adverse symptoms. The first to show symptoms—mild skin eruptions, later diagnosed as chloracne and conjunctivitis—were maintenance engineers, fitters, plumbers, and electricians. All these men worked with their bare hands, suggesting that factory surfaces had been contaminated. The regular factory employees always wore gloves and contracted chloracne weeks later. Between May and December 1968, 79 cases of skin disease were recorded. In every case the face of the worker involved was affected, those with a fair complexion being the most severely marked; but in workers with a more sallow complexion the condition was more prolonged.[69] Arms, the back of thighs and calves, and the sternum were also affected.

Skin lotions, antibiotics, and ultraviolet light were all used to treat the chloracne. New and rigorous hygiene conditions at the plant were introduced with workers required to shower at work, change clothes on-site, and leave dirty overalls for laundering by the company's cleaners. According to May,[69] this hygiene regime was effective, for the skin condition of the men considerably improved in the ensuing 4-6 months. Age and skin complexion were again evident as factors controlling the rate of improvement; it was the younger, fairer-skinned workers who made the most rapid progress.

This outbreak of chloracne between May and December 1968 again forced the plant's closure. In addition to the hygiene conditions referred to above, workers were now required to wear protective clothing before entering the building, access to which was through a specially designated decontamination room. To assess the extent of the contamination the standard wipe tests were carried out. Small pieces of cloth are literally wiped across surfaces to remove chemicals from all items thought to be contaminated. The cloths are thoroughly washed with suitable solvents and the chemicals of interest leach out. Further tests are then performed to identify the materials. In the case of chlorinated compounds the standard test is done involving the ear of rabbits.[71] These animals are particularly sensitive to the chemical, as was discovered in 1957 by one scientist studying chlorinated compounds. Dr. Karl Schultz, a German dermatologist, had been asked to identify the cause of an outbreak of chloracne in workers employed by the German company, Boehringer. Boehringer had itself had an accident in its trichlorophenol plant in 1953. Schultz's investigations eventually pinpointed dioxin as the chemical causing chloracne. Schultz found that minute quantities of dioxin would produce an inflammatory response in the rabbit's ear; larger amounts of the chemical would kill the animal in a very short time (see Chapter 5 for more details).

When Coalite's wipe and rabbit ear tests produced the same effects reported 15 years earlier by the German scientist, the evidence was irrefutable. Coalite's plant was obviously still contaminated and something had to be done about it. As far as Coalite was concerned the only sensible solution was to dispose of all the equipment directly damaged and seriously contaminated by the blast; it was buried 150 feet down the shaft of an open-cast coal mine.

Production of trichlorophenol resumed in 1969 in a new plant extensively modified as a result of the accident a year before. Modifications introduced included temperature control which was not automated, and each reactor was now fitted with five failsafe functions independent of the program control. These safety devices were designed to shut down the reactor automatically at any critical point and would, in theory, prevent an exothermic reaction from developing.

In March, during the construction of a new building at the Bolsover complex two workers developed persistent chloracne. The men were employed to install the new plant by a local contracting firm, Site Engineering Services of Chesterfield. Three months after they began work at Coalite, the characteristic chloracne blackheads appeared. A company investigation traced the source of the problem to a large metal vessel contaminated with dioxin in the trichlorophenol reactor explosion three years earlier. Coalite claimed that the vessel had been extensively cleaned by its employees before the site contractors handled it. But the cleaning was clearly inadequate, for not only did the son of Anthony Spillane—one of the workers involved—develop chloracne, but the wife of his co-worker did as well. According to Spillane, nobody informed him officially that there might be any hazard involved in working on the trichlorophenol plant.[72] Although he had noted that many of the regular Coalite employees were "spotted" and "pockmarked," it was only a month after starting work that a colleague informed him about the risk of contracting chloracne. Two months later he developed the skin complaint and shortly after this his 4-year-old son began to show symptoms, presumably as a result of contact with clothes of his father which were contaminated on-site. Both father and son recovered from the episode with treatment, but five years later in 1976 the child was still pockmarked.

The comments of one other Coalite worker who contracted chloracne are similarly revealing of the distressing effects of the disease; Ernest Taylor developed the skin disease in 1968, two weeks after starting work on the damaged reactor piping. Referring to the period when he had the skin disease, Taylor is quoted as saying: "It was a very worrying time. You could eat and sleep, but didn't dare go out in company because your face was in such a mess."[73] He added that none of the workers knew the cause of their skin complaint. "It was frightening" he said.[73] Taylor's main worry was that he might pass the chloracne on to his family. Fortunately this did not happen, but the persistence of his complaint is evident from the fact that skin ointments and steam baths did little to hasten the departure of the disease. When the blackheads did eventually disappear they were replaced by scars which took over two years to fade. As compensation for his injuries Taylor received several hundred pounds from Coalite. The other workers with chloracne were similarly compensated.

In addition to the two contract workers with chloracne, an additional 12 cases have been reported by Coalite since 1968. The total number of cases now exceeds 90. There is only one recorded death among the workforce so far. One former employee is said to have died from a coronary thrombosis.

The Coalite management had adopted a relatively low-key approach to the problems in its 2,4,5-trichlorophenol plant. A single event in the summer of 1976 changed this. On 10 July a trichlorophenol reactor near the Italian town of Seveso overheated, discharging its contents over the town and its environs. The attendant publicity worried all manufacturers producing the chlorinated phenol. All were concerned about the consequences of the accident and Coalite was no exception to this rule. The company was still in the business of producing trichlorophenol and reporters on the local Sheffield newspaper, *The Star*, had become interested in what was happening at the plant. The paper ran a series of articles on the company. At the time of the Seveso accident, Coalite's reactor had been shut down for the holiday period. Charles Needham, Managing Director of Coalite, was not impressed with *The Star*'s coverage. The paper, he said, had been most unfair to Coalite.[74]

But *The Star*'s articles worried the Coalite management. On 13 August 1976 the company took out quarter-page advertisements in some national newspapers including *The Times*, *The Guardian*, and the *Daily Telegraph*, in an attempt to reassure the public that there was nothing to worry about, the trichlorophenol process was safe. They stated that "The Coalite 2,4,5-trichlorophenol unit incorporates the most modern, automated, "failsafe" systems. As part of our normal annual holiday shut down arrangements the unit was already idle when the Seveso accident was first reported. Although it was ready to start up a week later, we appreciated that anxieties could arise and decided it should remain shut down. We shall NOT make any decision on reopening it until we have satisfied not just ourselves but also our workforce, the local authority and the Government's Health and Safety Executive that it is '110% safe.' "[75]

The statement went on in an optimistic vein to confirm that Coalite believed its specially constructed 2,4,5-trichlorophenol unit at Bolsover was ". . . safe against any occurrence such as that in Italy." This assertion was never put to the test. Coalite failed to convince the most crucial group—its workers—that the plant was safe and on 14 October 1976 Needham conceded, reluctantly, that the plant would close permanently. Announcing the decision in a terse statement, Needham said that closure would mean ". . . a loss of jobs, wasted capital investment, a loss of exports (valued at about £1 million) and an increase in imports."[76] Needham attributed blame for the closure on ". . . oversensational reports and comment following the accident at Seveso."[76]

The reason for Needham's irritation is not hard to find. Coalite is not the most communicative of firms. When *The Star* began its reports on the company in July 1976, they were based on the medical report on the Coalite workers written by the company doctor George May. Coalite insisted that it encouraged May to write the article. To then find itself in the dock, so to speak, because of its openness in the past, was, in Coalite's view, unfair.

For *The Star* there were two issues of interest. The first was discovering the site where contaminated material from the 1968 explosion had been dumped, and the second was the "human interest" stories on the Coalite workers. Needham was less than candid on the site of disposal. He refused to disclose the whereabouts of the site to many enquiring local councillors following *The Star*'s reports on the grounds that (a) it was not located within their particular council boundaries, and (b) that revealing the site would only increase public anxiety.

One councillor, however, was informed of the whereabouts of the site. David Bookbinder, Chairman of the Derbyshire County Council's Planning Committee, pronounced himself satisfied with the information he had received from Coalite. In his opinion, Coalite was taking the "socially right line"[77] in not revealing the information to all and sundry. Bookbinder refused to reveal the site himself, saying that it was for Coalite to make the announcement.

Coalite refused to provide the information. Addressing local councillors at Bolsover, Needham claimed that the contaminated plant was well out of harm's way and he insisted that the dioxin contaminating the equipment was harmless two years after burial.[78] Needham was guessing about the rate of degradation of dioxin—nobody knew it at that time. Experiments conducted in the United States and Italy would later show that the rate at which dioxin degrades in soil is variable and depends on the location. In the soils of Florida microorganisms will reduce the quantity of a given amount of dioxin by half in 220 days. After a further 220 days the concentration of dioxin would be further reduced to a quarter of its original value.[79] The 220 days required to lower dioxin concentration by a half is referred to as the chemical's half-life. The same measurement is used with radioactive chemicals. At Seveso, however, the half-life of dioxin is considered to be in excess of 10 years.[80] There is no information available about dioxin decay under the conditions in which Coalite's contaminated equipment was buried. Needham was correct about one thing though—the equipment was indeed safer underground where no one could come in contact with it.

Four years after Needham had reassured the councillors that the dioxin-contaminated equipment presented no health hazard, the issue surfaced again. Residents in villages in north Derbyshire began to ask questions about the actual burial site of the equipment. Many suspected that it was located in a waste tip and a field close to the village of Morton.[81] The residents were concerned because a recent decision by Derbyshire Council's planning committee had given the permission to go ahead with coal exploration in the area. The Council, however, has said that the residents have nothing to fear, for the contaminated equipment will not be disturbed by the new activity in the area. And once again, councillors have decided that it is not in the residents' interests to reveal the whereabouts of the site; in their opinion it would cause needless anxiety.[81] In a rare television interview, Needham repeated the message: "There was nothing to worry about," he told the viewers.[82] It seems unlikely that this situation will be allowed to continue. Some local councillors may yet call for an inquiry to discover the whereabouts of the contaminated equipment.

Another issue which remains unresolved is that of the disposal of the dioxin-contaminated waste from the routine manufacture of 2,4,5-trichlorophenol. This waste is usually heavily contaminated with dioxin and it should be treated with ultraviolet light or by incineration at high temperatures to destroy the chemical. Manufacturers of the trichlorophenol in the United States and Holland chose another method of disposal. The waste was dumped in the nearest tip or landfill, creating a considerable pollution problem which will be costly to amend. Coalite never revealed how it routinely disposed of dioxin-contaminated waste. Many feel that it would be helpful if it did.

Safety was a subject of no little significance for Coalite's workforce. As Ernest Taylor's earlier comments confirm, the workers resented not being fully informed of the dangers they faced on the trichlorophenol plant.[73] But the Coalite management was not, itself, fully aware of all the dangers in 1968. George May—also a victim of chloracne, which he contracted in the course of his examination of the Coalite workers—wrote to the Monsanto Company in the United States for information relating to a similar accident which it experienced in 1949; Monsanto never replied.[83] May did receive information from the German chemical company Boehringer detailing the problems it had faced following its own accident in 1953. Boehringer also informed May about Karl Schultz's work on chloracne.

Eight years later, more information would have done little to persuade Coalite workers that the trichlorophenol reactor was safe. A week before Needham announced that the plant was to close permanently Coalite tried, unsuccessfully, to recruit workers to man it. 150 fitters, their mates, and instrument technicians already working for the company were unwilling to be transferred to the plant.[84] In the months following the managing director's announcement on 14 October 1976, the workers previously engaged on the modified trichlorophenol plant were redeployed to other sections of the Bolsover complex. The move was not made, however, before a dispute with the principal union involved—the Transport and General Workers Union—forced the issue.[85,86]

The publicity which Coalite attracted following the Seveso accident was only one of several factors which brought about its closure. Financial considerations were probably uppermost in the minds of the Coalite directors as Needham revealed in his statement announcing the closure. According to Needham, the trichlorophenol plant—rebuilt in 1969, the year after the accident—had made a loss in 1975 and had not repaid capital investment. In addition the Health and Safety Executive (HSE) had insisted on additional safety features being incorporated before the plant could reopen.[87] These factors, together with the hostility of the union, and the outrage of the local Labour member of Parliament, Mr. Dennis Skinner, over the lack of information emanating from the company,[87] provided the extra push required to close the plant. Commenting on the reasons for the closure, the assistant managing director Mr. R. P. Marshall said: "No one reason was overwhelming. It was the totality. And the desire to be good neighbors with people in the neighborhood and the local authorities."[88]

Marshall's comments appear a little disingenuous in the light of subsequent developments at Coalite. Shortly after the Seveso accident, the HSE insisted that Coalite carry out additional tests to assess the health of its workforce. George May was assigned the task of carrying out this surveillance. In consultation with the HSE it was agreed to test for deficiencies in the immune system and to look for any liver or blood lipid (fat) disorders.

For the program, May selected 121 individuals, 41 of whom were known to have been exposed to dioxin and who showed signs of chloracne, 54 of whom might have been exposed to the chemical, and a supposed control group of 31. The control group, as a Coalite report was later to show, was not an age- and occupation-matched population. Instead it was an inhomogeneous group, the numbers of which had been bolstered by the inclusion of Coalite management staff undergoing a regular lipid screen at the time.

In addition to the tests agreed with the HSE, Coalite undertook to carry out a more extensive series of tests after consultation with an outside contractor. The contractor, another industrial company, was anxious to obtain as much information as possible about the effects of dioxin on human health. It agreed to underwrite the costs of this more extensive research program, which would include looking for genetic damage in the dioxin-exposed workers, as well as a more comprehensive assessment of the immune system than had been initially suggested.

Two Sheffield University staff were commissioned to do part of the research; Dr. Eric Blank, Reader in Medical Genetics, was asked to make the assessment of chromosome damage, and Dr. Anthony Ward, Senior Lecturer in Immunology, was to assess the immune system. Dr. Jenny Martin, a consultant chemical pathologist at the nearby Chesterfield Royal Hospital, agreed to measure blood lipids and liver function.

An increase in chromosome irregularity of cells—such as lymphocytes in culture—is widely regarded as evidence of exposure to chemicals which are mutagenic, and potentially carcinogenic. Dioxin has been shown to be both a mutagen[89] and a carcinogen in animals[90] (see also Chapter 2 on Toxicology). However, Blank was unable to observe any chromosome damage—measured by variations in the frequency of chromatid–chromosome gaps and breaks or in the frequency of sister chromatid exchanges—which could be attributed to exposure to dioxin.[91]

Ward, on the other hand, did observe dioxin-related damage. He reported that the dioxin-exposed group showed a significant increase in the proportion of cases with reduced levels of the two immunoglobulins, IgD and IgM.[92] The results, said Ward, suggested that the dioxin-exposed group of workers had suffered from reduced immune capability and that their "short-term immunological memory" had been impaired. Workers in the category of "possible dioxin exposure" showed changes which were intermediate between the control and the main study group.[92]

There were also changes in liver function and blood lipids which were related to dioxin exposure, according to Dr. Jenny Martin. She said the results showed that the dioxin-exposed group had a greater incidence of impaired liver function as measured by the enzyme gamma-glutamyl transpeptidase. Furthermore, when the results for serum cholesterol, triglyceride, and high-density lipoprotein were subjected to multivariant analysis, they showed a significant difference between the dioxin-exposed group and the control. In the dioxin-exposed group, levels of serum cholesterol and triglyceride were higher and high-density lipoprotein lower than in the controls.[93] These are factors commonly held to imply an increased risk of developing cardiovascular disease.

Coalite informed Martin some time after she had completed her work that it did not wish to have the information published. She was also informed of the nature of the control group used for the study. Realizing that the study had been devalued by Coalite's choosing to include management staff in the control population instead of restricting it to chemical workers alone—the recognized practice in this type of study—Martin arranged a second study.

The blood chemistry of eight Coalite workers suffering from chloracne was compared with a matched control group. This second study was carried out without the involvement of Coalite and the results were published in the medical journal *The Lancet*[94] in February 1979. As in the first study, the results showed increased serum cholesterol and reduced serum high-density lipoprotein in the dioxin-exposed group. The differences were considerably more marked than in the original, larger study. However, they were not statistically significant, a point which Martin notes, but says is simply due to the small number of subjects involved.

Events now took a very different course. Shortly after her letter was published in *The Lancet*, Martin's house was broken into and the detailed medical records of the eight Coalite subjects were removed from her filing cabinet. Martin reported the theft to the local police constabulary, but as she offered no opinions as to why anyone would want to steal this information the police investigation never got off the ground. The police have since said that the case is not closed, but as there were no clues about the motives for the theft, it was unlikely that it would ever be solved.[95]

Understandably, Martin was very distressed when she discovered the theft of the material for the second survey—the only items removed from her house. She had no duplicate copy so the loss of the medical reports prevented her from publishing further details on the group of workers she had examined. An abbreviated form of Martin's first study was given to the Association of Scientific, Technical and Managerial Staffs (ASTMS), one of the trade unions involved at Coalite, by the company management. The abbreviated report contained the results of the study recommended by the HSE. It was quite different from Martin's original document. Coalite's brief report says there were no differences in the blood parameters between the dioxin-exposed and control groups of workers.[96] Martin, in her full report, which she submitted to the company, says there were differences.

The most serious aspect of the affair, however, was the position of the Health and Safety Executive. The Executive had not been able to obtain the full medical reports from Coalite. Lawyers within the Executive said that the provisions of the 1974 Health and Safety at Work Act (HSW Act), which established the HSE, were not sweeping enough to force Coalite to hand over this information. The Act only permitted the Executive to demand from companies data relating to a specific chemical and then only if the company were still producing it. Coalite, however, had ceased its production of 2,4,5-trichlorophenol in 1976. Thus, said the lawyers, the Act did not apply.

Employees of the HSE had seen Coalite's abbreviated medical report and were, therefore, not unduly worried about the health of the company's workforce. In March 1980, this position changed. In an article in the 6 March issue of the scientific journal *Nature*, Coalite's reluctance to release its medical reports was revealed.[95] The journal had been given a copy of Dr. Martin's study and duly reported its contents, together with a criticism of Coalite's version of events, and an account of the burglary.

The report in *Nature* attracted wider media publicity. Coalite at first denied that it had issued a biased version of Martin's report,[97] and the company's managing director Charles Needham, denied all knowledge of the full medical reports, insisting that the company had done nothing to be ashamed of.[98]

But the publicity had had its effect. Two weeks after the situation was first reported Coalite handed over both the abbreviated and the three full medical reports to the HSE for evaluation.[99] The Executive had asked for the documents and Coalite decided to comply with the request. There was no question of the law being invoked to secure their release. At the time the documents exchanged hands the HSE was still of the opinion that it lacked the necessary powers to compel Coalite to part with the information. Some weeks later, however, the HSE announced that following further scrutiny of the 1974 HSW Act it was found that the necessary powers were available to compel companies to release any information relevant to the Executive's work.[100] Privately, however, epidemiologists in the HSE admitted that the latest interpretation would probably have to be confirmed by a test case in court.

On 1 May 1980 the Executive issued its own evaluation of the Coalite documents. According to the Executive's Employment Medical Advisory Service (EMAS) the abbreviated report was of little value, and had been poorly prepared. EMAS criticized the composition of the three groups used in Coalite's medical program, stating that they were ill matched and that in consequence it would be improper to extrapolate the findings of the survey in any general way. The Executive noted that there were no signs of illness in the workers which could be related to their exposure to dioxin, but added that the significance of the biochemical results (on liver function and blood lipids) and immunological findings could only be evaluated in the longer term. To assist this evaluation, EMAS recommended that the workers exposed to dioxin and still employed by the company should undergo relevant medical tests every 3–5 years.[101]

But Coalite did not wait this long to repeat the tests. A few months after EMAS published its report, the company offered to reexamine the men included in the original survey. Although all were offered the test, Dr. George May, Coalite's doctor, said that "not everybody had availed themselves of the opportunity."[102] In the event, only 29 of the original sample of 41 workers known to have been exposed to dioxin were reexamined. The examination was, however, only partial; measurements to test the function of the immune system were not repeated.

The most significant omission from the studies so far is the absence of medical information on the majority of workers who were exposed to dioxin at Coalite. An estimated 90 individuals were affected but only 41 were included in the company's 1977/1978 survey. Attempts have been made by independent physicians to perform a comprehensive series of tests on all of the men who were affected, but these have apparently been rebuffed by Coalite. Documents made available to the scientific journal *Nature* showed that Coalite would not agree to any further investigations of its workforce if the physicians concerned insisted on the right to publish their findings. Faced with this response from the company at least one physician withdrew his offer of help.[102] Coalite's insistence on secrecy had other serious repercussions. The sponsor of the 1977/1978 survey indicated that further support would not be forthcoming if Coalite persisted with this policy.[102]

Without this source of income Coalite may well decide that it is not in its interest to mount further investigations. As British law stands the company may have the last word. It is still not clear whether the necessary powers exist under the 1974 Health and Safety at Work Act which would compel Coalite to undertake further investigations of its workforce.[102]

Evidence compiled by Coalite for the 1977/1978 medical survey indicated that there had been no excessive mortality rate among its dioxin-exposed workers. The company had traced 79 of the 90 individuals affected and only one death—from a coronary thrombosis—had been noted in this group. The company had not provided information on the other 11 individuals. All those affected by dioxin at Coalite were exposed to the chemical either during or after 1968. A realistic mortality picture of this population cannot be assembled for many years to come.

But criticisms of Coalite's attitude to the health of its workforce persists. A report in *The Star* on 27 June 1980 claimed that many of Coalite's former employees exposed to dioxin had been traced by the paper. Of the 26 it had interviewed, 18 had complained of heart problems.[103] Martin's work had suggested that these men may have a greater risk of developing heart complaints. However, in the absence of a proper epidemiological study, no firm conclusion can be drawn about the incidence of heart abnormalities in these men and nothing can be inferred about the cause of the complaints.

The repercussions of the Coalite affair are likely to continue for some time and the company is not alone in having laid itself open to criticism. The actions of the HSE over Coalite have been brought into question by ASTMS, the union most involved with the situation.[104] It is the union's view that the Executive was caught out by Coalite. And to compound the problem, the union claims that the HSE failed to keep it informed of developments at the firm as it had promised.

For the HSE, Coalite has been a chastening experience and it is unlikely to be caught so unprepared in future. The developments at Coalite have shown that the legislation to secure a safer working environment in Britain is still far from perfect and it is clear that some action is necessary to remedy this situation. For the dioxin-exposed workers at Coalite the remedy they need is assurance that their health will not be impaired in the long term.

Soviet Union, 1970

There is only one significant report on dioxin exposure of workers to come from the Soviet Union.[105] Prepared by Drs. K. A. Telegina and L. I. Bikbulatova of the Department of Occupational Skin Diseases of the Ufa Research Institute of Hygiene and Industrial Medicine, the report describes the symptoms of 128 workers engaged in the manufacture of the herbicide 2,4,5-T, all of whom had symptoms of occupational skin disease.

In their paper Telegina and Bikbulatova refer to the fact that studies have been made on the risks users of 2,4,5-T might encounter. They also refer to observations of skin disease associated with the use of chlorinated hydrocarbons, chlorinated naphthalenes and chlorinated diphenyls, three compounds known to cause chloracne.[106]

The Soviet workers involved in the manufacture of 2,4,5-T were initially examined a year after production of the herbicide was begun, and a second examination was made a year and a half later. Only 83 out of the 128 received extensive clinical examinations. Of these, 13 had severe chloracne, 24 had "moderate" signs of skin disease, 32 had minimal evidence of it, and 14 had no clinical evidence of chloracne—their skin complaint presumably being some form of dermatitis. The workers most severely affected tended to be engineers involved in the supervision of the industrial process. Chloracne was evident in this group far earlier than in the others. In addition to the skin disease, the workers complained of symptoms as diverse as headache, somnolence, stomach pains, nausea, heartburn, irritability, fatigue, loss of memory, pain in the heart, loins and joints, sleeplessness, a tendency to perspire, and a lowered sexual potency.

As with the Coalite workers chloracne was most evident on the face; the ears, neck, back, chest, shoulders, and genitals also had the characteristic blackheads and skin eruptions of chloracne. Disorders of the nervous system were observed in 18 workers, and 17 had chronic tonsillitis. The authors report a high incidence of ear, nose, and throat disorders in the workers generally. Their conclusion is that the involvement of the upper respiratory tract is no coincidence and is presumably related to the irritant action of the various vapors given off during the manufacturing process as well as the drying effects of the temperature gradients.

Results of clinical tests on the workers are less clear-cut. Although greater changes in the blood white cell counts were seen in some of the 13 severely affected, most of the values fell within normal limits. Similarly a tendency to hypercholesterolemia (high blood cholesterol) in the 13 was not statistically significant and the mean cholesterol value of all four groups was well within the age adjusted norm, at 160–180 mg/100 ml (milligrams per milliliter [4.1–4.7 mmol/liter (millimoles per liter)]. These values are at the lower end of the normal range for this age group (160–220 mg/100 ml), according to the authors.

Liver function tests on all the workers, produced results suggesting that the livers "antitoxic action" and lipid metabolism had been affected; however, pigment formation and carbohydrate and protein metabolism were not impaired.

Irradiation of the skin with ultraviolet light is a well-known treatment for many skin disorders. To avoid burning, care in the use of uv lamps is essential. In the course of their treatment regimes the authors noted a decreasing sensitivity to uv light with increasing severity of the skin disease.

Concluding their report, Telegina and Bikbulatova say that regression of the chloracne was observed following a course of treatment of antibiotics, vitamins, and skin lotions but that the decisive factor in the elimination of the complaint was the decision of the factory management to cease production of 2,4,5-T and change to an alternative herbicide.

Missouri, U.S.A., 1971

An environmental incident involving dioxin, and exceeding that at Seveso in its severity, occurred in eastern Missouri in 1971.[107] The Bliss Waste Oil Company was commissioned in early 1971 to dispose of dioxin-contaminated waste from a trichlorophenol reactor. The waste, amounting to several thousands of gallons of residue and containing at least 12 lb of dioxin—the result of many operational runs of the reactor—was collected by Bliss and was in due course sprayed onto several horse arenas where the firm had been hired to assist with dust suppression. Bliss did not realize that the waste represented a potential hazard, and did not treat it as such.

The credit for identifying dioxin in the waste oil belongs to Dr. Renate Kimbrough, M.D., a pathologist, employed at the Department of Health, Education and Welfare's (now the Department of Health and Human Services) Center for Disease Control (CDC) in Atlanta, Georgia. Her involvement began with the arrival of specimens at CDC of animals which had died in the precincts of the horse arenas. Animal deaths began to occur in the days following the spraying of the first arena on 26 May 1971. Hundreds of birds died in and around the arena in the first week. Cats and dogs were the next to succumb and in the following weeks eleven cats and dogs died. Horses previously in good health became ill after exercising in the arena and began to show signs of what was later shown to be dioxin poisoning. Four days after the spraying the first horse began to show signs of intestinal colic; it died three weeks later. Despite excavation of the area soil in October 1971 and April 1972 and its replacement by fresh earth, horses continued to die until January 1974. Of 85 horses which exercised on the contaminated soil in the summer of 1971, 58 became ill, 43 of which subsequently died[108] after exhibiting distressing symptoms including anorexia, mouth and skin ulcers, fever, ataxia (unsteady gait), and foaming at the mouth. In horses that were terminally ill, the length of illness varied from 4 to 132 weeks depending on the extent of their exposure to the contaminated soil. Contact with the contaminated soil in the arena caused 26 mares to abort; those foals carried to term were presumably exposed in utera, for many either died at birth or shortly thereafter.

Accounts differ as to the number of horse arenas sprayed with trichlorophenol waste oil. Kimbrough[107] refers to three arenas; Dr. Barry Commoner at Washington University's Center for the Biology of Natural Systems in St. Louis says there were four.[108] The sequence of events described by both is, however, similar. The salvage oil company sprayed additional arenas with trichlorophenol waste a few weeks after the first spraying at the end of May. Within days birds began to die, then cats and dogs, and finally horses. Twenty-five horses permanently stabled in a second arena died. In a third—an open riding arena—seven animals died after training on the oil-soaked soil. Thirty others were severely ill but fortunately did not die.

But it was not only animals that were exposed to the oil waste in the arena; children, the owners of the stables, and spectators attending riding shows were also in contact with the oil. The severity of their symptoms, however, was far less than that seen in the animals. Indeed, in view of the quantity of dioxin deposited in the horse arenas the low level of illness reported is remarkable. To assess the amount of dioxin in the arenas, samples of trichlorophenol waste oil remaining in the tanks of the company producing the trichlorophenol were later analyzed for the dioxin content by Kimbrough and her colleagues. High values of 306–356 ppm were recorded by her. It is known that at least 4000 gallons of waste oil were sprayed onto two arenas.[108] How much was put onto the others is not known. One kilogram of waste oil would contain 306–356 mg (milligrams) of dioxin. Four thousand gallons of oil, weighing 18,200 kg would contain 5569–6479 g (12–14.2 lb) of dioxin. The quantity of dioxin said by the Italian Authorities to have been released at Seveso is about 250 g, or about 1/2 lb. The dioxin at Seveso was discharged over an area many times greater than the horse arenas. From the figures available it can be calculated that the amount of dioxin on the Missouri soils was 300 mg/m^2 (milligrams per square meter),[108] whereas that in the heavily polluted Zone A at Seveso was 0.5–1.0 mg/m^2. The Missouri figures represent a level of contamination 300–600 times that of Seveso.

It is all the more surprising, therefore, to discover the number of people officially recorded as suffering symptoms of dioxin poisoning. One account refers to ten people[109] affected by the chemical, but detailed information is only available for seven. Of these three were adults and four were children under ten years of age. Both owners of the first arena (A) were affected shortly after the initial spraying with contaminated waste oil. One co-owner of arena A acted as the ringmaster for a horse show held four days after the spraying; he developed a severe headache and nausea while in the arena. The other co-owner worked in an office adjacent to the arena; she developed severe headaches with occasional nausea and diarrhea and was forced to leave the office after two weeks. The third adult affected was a vet called in to treat horses at the third arena after spraying. He is said to have been exposed to dioxin while collecting soil samples from the arena for analysis. He is recorded as having developed the persistent and particularly unpleasant skin disease chloracne and to have suffered from this complaint for one year at least.[108]

Those most seriously affected in the Missouri incident were the children. A ten-year-old girl rode horses in arena A two or three times a week in June and July 1971. She became listless, had nosebleeds and headaches, and occasional abdominal pain and diarrhea. Her symptoms are said to have disappeared in August 1971 when she stopped using the arena.

A second child, the six-year-old sister of the first girl, was also affected in Arena A. The younger girl played daily in the contaminated soil of the arena in the summer months of June and July 1971. Like her sister she too suffered from nosebleeds and diarrhea and became listless. On 21 August 1971, when her symptoms appeared to be worsening, she was admitted to a hospital. The girl's complaints included difficult and painful urination. A medical examination confirmed inflammation of the bladder wall. Liver function tests proved normal. Her removal from the contaminated arena meant that she was no longer in daily contact with the dioxin. Her acute bladder symptoms subsided and, according to one report,[110] did not recur. A further report states that it took several months for the child to respond to treatment and that in the course of her illness she lost a staggering 50% of her body weight.[108] Another report on the girl was made $5\frac{1}{2}$ years later in the wake of the accident at Seveso. The same doctors who had examined her previously reported that they could detect no abnormal clinical symptoms.[110] The girl, they said, a victim of long-term exposure to dioxin, had grown normally, both in height and weight, in the intervening period between her first examination in 1971 and the second in 1977.

Two three-year-old boys make up the total of seven people exposed to dioxin. Both played in the contaminated soil of the third horse arena. Within six weeks of the first contact with the oil-soaked soil, both had the unmistakable signs of chloracne. In the following months the skin condition increased in severity before gradually subsiding a year later.[107]

Many people are certain to have been spectators at shows held in the contaminated horse arenas in the summer of 1971. As such they would have been exposed to the dioxin-contaminated dust churned up by the horses they were watching. It is impossible to say how many people were exposed to the chemical; there are no suitable records available. But it was not only the summer of 1971 which presented a hazard from dioxin-contaminated soil. It is known, for example, that the contaminated soil was present in the first arena (A) for five months before a 6–8-inch top layer was removed. A further 6-in. was removed five months later following the death of a cat which frequented the arena. The first layer of earth removed is reported to have been used as a "fill" in the construction of a new road.[111] In the case of a second arena sprayed with trichlorophenol waste in July 1971, the top soil was not removed for 21 months. When excavation took place a layer 12 inches deep was removed and used as fill in the yards of two eastern Missouri houses. The toxic fill is said to have killed four trees.[108] In the third arena (C) where the two three-year-old boys were exposed to dioxin, the contaminated soil remained untouched for over a year before a 12-inch layer was removed and placed in a specially prepared landfill, at 70 feet below ground level.[112]

Documenting what happened to the topsoil proved to be a far easier exercise than that of actually identifying the chemical causing the animal deaths. Finding the source of the contaminated waste oil was a similarly daunting task. Dr. Kimbrough and her colleagues at CDC were involved in a hunt for the actual producer of the waste. Kimbrough herself mentions no names in her description of the Missouri episode.[107] However, a letter from Dr. David Spencer, Assistant Surgeon General in the then Department of Health, Education and Welfare,[113] to Senator Philip Hart (Chairman of the Senate Sub-Committee on the Environment) is less reticent. Spencer's letter, written in response to an enquiry from Senator Hart for more details concerning the contaminated arenas, gives a step-by-step report of the detective work necessary to identify the source of the contaminated waste.

According to Spencer[113] analysis of the contaminated soil in eastern Missouri had revealed the presence of dioxin. The pattern and concentration of the chemicals in the soil suggested industrial waste. However, an investigation of the records of the Bliss waste oil company failed to reveal direct evidence of any such waste which might have contained dioxin. Thwarted by this approach, the investigators' next step was to review all companies producing 2,4,5-trichlorophenol and the herbicide 2,4,5-T, prior to 1972. The company responsible was finally identified as Northeastern Pharmaceutical and Chemical Company Inc. (NEPACCO) of Stamford, Connecticut. But its identification was by no means a simple task. The trichlorophenol reactor in southern Missouri was first used by Hoffman Taft Inc. (now Syntex Agribusiness, Inc.) for producing 2,4,5-trichlorophenol for conversion into the herbicide 2,4,5-T. In February 1964 Hoffman Taft ceased production of 2,4,5-T, but in December of that year it leased the trichlorophenol plant to NEPACCO. Between April 1970 and December 1971 NEPACCO produced trichlorophenol for conversion to the bacteriacide hexachlorophene. At the end of each production run, trichlorophenol was distilled off for purification and the distillate residue was stored in a large tank. According to Spencer a preliminary analysis of residue in the tank gave dioxin valves in the 1000-ppm range; later measurements suggested that it was 306–356 ppm.

Disposal of the waste was initially carried out by Rollins Purle Inc. (now Rollins Environmental Service Company) of Baton Rouge, Louisiana. In 1971 NEPACCO was in financial difficulties and it commissioned another contractor, Independent Petrochemical Company, Inc., St. Louis, to handle its waste. Independent Petrochemical in turn subcontracted the waste disposal work to the Bliss Waste Oil Company. Bliss is reported to have removed between 10,000[112] and 18,000[107] gallons of waste from NEPACCO's tanks between February and October 1971.

At no time during the disposal of the trichlorophenol waste did NEPACCO inform the haulage contractors that they were dealing with toxic material. Rollins-Purle were told only that their services were required because NEPACCO lacked storage space for its waste. Independent Petrochemical and Bliss were also kept in the dark; neither was told about the hazards involved in handling the waste. Bliss routinely collected oil from various sources for reprocessing. Used oil from service stations, trucking companies, and airports was stored in four 24,000-gallon holding tanks before being sold to oil refineries for reprocessing. During storage the oil clarified leaving sediment to settle as sludge in the tank bottom. This formed about 10% of the oil in the tanks, was not acceptable for reprocessing, but was ideal for use as a dust suppressant.[114] This was what happened to the waste from the NEPACCO plant. Between February and October 1971, six loads of trichlorophenol waste—about 18,000 gallons according to Kimbrough—were collected from NEPACCO and stored in one of the tanks at the Bliss depot. It was the sludge from this tank, toxic in the extreme, which was removed and sprayed onto the horse arenas in eastern Missouri.

In an interesting anecdote Kimbrough[107] reports that on 21 May 1971, on the third haul from NEPACCO to Bliss, the salvage oil company's truck driver received a traffic citation for overloading. To reduce weight, a portion of the trichlorophenol waste was dumped on a farm road owned by the director of Bliss Waste Oil. Within two weeks 70 of his chickens, which had rummaged in the sprayed soil, had died. In 1974, soil samples collected from the farm road still had measurable concentrations of dioxin.

By late 1974, when dioxin in the toxic soil had been removed, the outstanding problem for Missouri's Health Department was the disposal of 4600 gallons of trichlorophenol waste in NEPACCO's storage tank. The waste had remained undisturbed since December 1971 when the company had ceased to produce trichlorophenol owing to the collapse in the hexachlorophene market. (The collapse followed the announcement that the bacteriacide hexachlorophene was neurotoxic and that it had caused nerve damage in French infants in 1972 sufficient to kill 36 children and seriously injure a further 145. The hexachlorophene had been used in talcum powder at a concentration of 6.35%. Concentrations of hexachlorophene should never have exceeded 0.5%, the reported upper limit.) All attempts to interest commercial companies in the disposal of the trichlorophenol waste were without success. No one was prepared to handle the material. In desperation the Missouri Health Department turned to the Environmental Protection Agency for help. It asked the EPA to grant an emergency permit for the waste to be disposed of on board the incinerator ship Vulcanus[115] (used to incinerate thousands of gallons of the herbicide Agent Orange left over from defoliation missions in Vietnam). But at first the EPA did not comply with the request. According to Dr. Kimbrough the Missouri Health Department also appealed to other agencies including Federal Government departments, but to no avail; none came forward with an offer of help.[116] A senior EPA official told the author, however, that the State of Missouri was offered facilities to have the waste incinerated on the Vulcanus. According to the official, the Vulcanus could have collected the waste from the docks at the Missouri town of Gulfport. The state authorities, the official claimed, "dropped the problem," preferring to leave the waste in storage rather than having the material crossing different state lines.[117]

The dioxin-contaminated waste is still in the storage tank on the original NEPACCO site, but fortunately it is now less toxic than it was. The 4600 gallon trichlorophenol residue contained some 350 ppm of dioxin—about 16 lb of the chemical. At the time of writing, some 90% of this has now been destroyed following treatment with ultraviolet light (a recognized method of disposing of the chemical).[118] As for the remainder of the dioxin, this is in a thick sludge which will require solubilizing before it too can be treated in the same way.

Unfortunately that is not the end of the matter. It appears that NEPACCO sold dioxin-contaminated waste to more than one customer. The incidents in which the Bliss Waste Oil Company sprayed waste on horse arenas in the State of Missouri have already been mentioned. But NEPACCO's other customers have sprayed the waste on at least seven additional landfills, one of which has been used for building an amenity center. Appropriately perhaps one EPA official described this catalogue of incidents as a "disaster."[118]

There can be no approbation for the manner in which the Missouri incident was handled. Indeed, it is perhaps a little ironic that many Americans considered the Italian authorities to be inept in the handling of the Seveso accident. The charge is certainly true; the Italian authorities did make many mistakes at Seveso. But in Missouri, despite obvious signs that something was seriously wrong, no action was taken for three years, not until the painstaking detective work by Kimbrough and her colleagues finally resulted in the identification of dioxin. By then it was too late to trace those people who, as spectators for example, may have been in contact with the chemical, in concentrations many times greater than that which was measured at Seveso.

Other Incidents

All manufacturers of 2,4,5-trichlorophenol, the starting point for the herbicide 2,4,5-T, have suffered accidents in which workers have been affected by exposure to dioxin. The incidents vary in the severity of their effects, and for some information concerning the consequences of the incidents is not readily available.

Diamond Alkali: Newark, New Jersey

Although treated briefly, all the incidents considered in this section have their own importance. All yield evidence vital to the assessment of the risk to humans posed by exposure to dioxin.

Diamond Alkali, a major supplier of 2,4,5-T in the United States, agreed to a health survey in February 1969, examining workers involved in the production of the herbicides 2,4,5-T and the closely related 2,4-D (2,4-dichlorophenoxyacetic acid). Seventy-eight male workers were examined; 48 had some evidence of acne.[119] Thirteen of the 48 had moderate to severe skin lesions, the severity of the lesions correlating with the presence of scarring, excessive pigmentation, abnormal hair growth, and complaints about eye irritation.

Six years earlier, 29 workers at the same plant had been medically examined. Many of the group had chloracne, and 11 workers had an abnormal excretion in the urine of uroporphyrins (pigments involved in the production of the blood protein hemoglobin).[120] The workers with this abnormal urine profile—termed *uroporphyrinuria*—were also noted to have fragile skin, and the characteristic skin eruptions of chloracne. Six years later, when the men were reexamined, they still had chloracne.

Spolana—Czechoslovakia

Some of the most serious cases of dioxin poisoning are to be found among workers involved in the production of 2,4,5-T at the Spolana chemical plant in Czechoslovakia. Some 400 workers were involved in the process and approximately 80 men are known to have been affected during occupational exposure to dioxin between 1965 and 1968. Fifty-five have been traced and medically examined.[121,122] Chloracne has healed in about a fifth of the patients but eight men still have a severe form of the skin disease with large cysts and abscesses.[123] In those cases where these lesions have healed, severe scarring has left many of the group permanently disfigured.[124]

According to one report of the accident, the first symptoms of intoxication which occurred at the time of exposure were gradual. These included chloracne, feeling ill, fatigue, weakness in the lower limbs, and pain on the right side beneath the rib cage.[124] Ten patients, however, only developed symptoms several months after their exposure to dioxin had ceased.

According to doctors at the Faculty of General Medicine at Prague's Charles University, the progression of the illness was variable. Some patients only showed mild symptoms initially but their condition became worse over the following 3–4 years. In others who at first appeared to have no detectable damage to organs, such as the liver, there was a sudden deterioration in their condition. Factors such as stress or unusual physical exertion are thought to have precipitated these crises.[124] Some specific changes were noted in the men. Over half had abnormally high blood cholesterol and phospholipid levels. Many had disordered protein and sugar metabolism, and some still have these symptoms. In others, poisoning was evident in the excessive excretion of uroporphyrins and mild impairment of liver function was noted in a fifth of the cases.

About a quarter of the men exhibited damage to the peripheral nervous system immediately after their exposure to dioxin, and in most their condition deteriorated in the following 3–4 years. But this was not true for all subjects. In some nerve damage was only evident months after their exposure. Fortunately the rate of deterioration slowed down and after four years no further impairment of the nervous system was evident. Apparently in three cases there was even some improvement.[124]

Ten years after the accident the blood fat disorders appear to have corrected themselves—the average blood lipid levels of 44 of the dioxin-exposed workers were similar to a comparable control group. Improvements in liver function have also been observed in those cases where damage was initially evident.

Psychiatric problems resulting from the disfigurement caused by chloracne were common. Anxiety and depression caused marital discord for some, and many young workers despaired of forming intimate personal relationships because of their disfigurement. Fortunately as the clinical condition of the men improved and their skin disfigurement became less obvious, these psychiatric problems were gradually resolved.

According to Professor L. Jirasek also at Charles University, the dermatologist who examined the Spolana workers, hygiene conditions at the plant were poor.[123] Dust from chemicals was a persistent problem, and there was a great deal of it in the plant, much of it from chlorinated compounds other than 2,4,5-T. A review of the hygiene problem in the plant convinced the management that conditions were so bad that the operation ought to be abandoned. Production stopped in 1968.

Chronic intoxication with dioxin does not appear to have led to an increase in the number of malformations among workers' children. In the intervening years since the accident, 18 children were born to the families of workers affected by moderate to severe dioxin poisoning. None of the children had any birth defects. As for spontaneous abortions two were recorded in 20 pregnancies, an incidence of 10%. Worldwide, some 15%–20% of pregnancies are said to end in spontaneous abortions.[125]

Twelve years after the accident, six of the 55 workers have died.[124] Two men were killed in traffic accidents, one died from liver cirrhosis, another from an atypical atherosclerosis, and two died of brochogenic lung carcinoma. The two men who had lung cancer died two and three years, respectively, after their exposure to dioxin. This short latency period between exposure to dioxin and the development of the tumors suggests that the chemical is unlikely to have caused the disease.

Boehringer

Between 1954 and 1955, 24 employees at the Hamburg site of Boehringer, one of West Germany's giant chemical companies, developed chloracne. All were employed in the production, handling or testing of 2,4,5-trichlorophenol, in either the mixing unit of the chemical plant, or in the laboratory. The men were recruited to the plant in 1953 and the early months of 1954. During 1954 the first symptoms of chloracne appeared on the face, arms, and nape of the neck of several workers. The numbers affected increased as the year progressed and by 1955, 24 had the skin lesions.

Additional medical complaints included loss of appetite, feeling of gastric pressure, intolerance of certain foods, nausea, weakness in the arms, and legs, general fatigue, loss of libido, and, in two cases, changes in heart rhythm reflected in alterations of the normal electrocardiograph (ECG) patterns. Five men had conjunctivitis and disturbances of liver function; liver damage was noted in eight others.[126]

Twenty years later, in 1976, 11 of the original 24 workers were reexamined. Many were continuing to suffer from their earlier complaints. In seven of the 11, nausea and intolerance of heavy fatty food was still common. Six men complained of alcohol intolerance. Although conjunctivitis had disappeared, chloracne was still clearly visible in most of the group. Neurological problems were still severe in six of the workers and, in consequence, they are receiving disablement pensions of between 20% and 50% of their normal salary. Two of the original group have died; arteriosclerosis and "nonnatural causes" are given, respectively, as the causes of death.[126]

Rhone-Poulenc: Pont de Claix—France

A French manufacturer of 2,4,5-T has also had industrial hygiene problems in the chemical plant used for the production of the herbicide's basic raw material.[127,128] Case reports on 17 workers exposed to dioxin following an explosion in 1956 in the 2,4,5-trichlorophenol reactor describe the appearance of chloracne in the affected men, conjunctivitis in 50% of the cases, and general digestive complaints consistent with anorexia and flatulence after eating.[129]

Although suggestions were put forward for measures to prevent a recurrence of chloracne at the site, the problem did reappear after a second explosion in 1966. Twenty-one men are said to have been exposed in this second incident. Poor industrial hygiene in the plant has further increased the numbers affected by dioxin exposure. Details, as yet unpublished and confidential, claim that around a hundred workers in all have been affected by dioxin contamination at Rhone-Poulenc between the year 1953 and 1970. However, this higher figure is as yet unconfirmed. As far as further studies are concerned there is no information available about potential mortality studies on this group of men.

Dow Chemical

Like some of its European counterparts, Dow only experienced serious problems with dioxin contamination some years after it started to manufacture 2,4,5-T. Production of the herbicide began at Dow in 1950[130] and continued until September 1979. This was the year in which the U.S. Environmental Protection Agency took action against 2,4,5-T, banning its use in forestry and rights of way. The ban was enforced in March 1979 on the grounds that a study conducted by the EPA showed that the use of the herbicide was associated with an increase in spontaneous abortions in the State of Oregon.[131] Several independent studies have since suggested that the EPA study had serious flaws, and that consequently its conclusions are questionable.[132,133]

The Agency believes, however, that 2,4,5-T is dangerous. It is seeking to change the partial restriction to a total ban on all uses of the herbicide in the United States.[134] However it may be prepared to compromise, and providing dioxin is removed from the herbicide the Agency may agree to the continued use of 2,4,5-T.

Until it ceased production in 1979, Dow was the largest company in the world manufacturing 2,4,5-T. The years 1964–1969 were profitable for all U.S. manufacturers of the herbicide; the U.S. Army was using 2,4,5-T liberally in Vietnam as part of its defoliation program, and chemical manufacturers had an ever-ready market for their product. Dow had a firm contract to supply herbicides to the U.S. forces. The company had few reservations about this contractual arrangement at the time. Only later did it have second thoughts. In November 1979 a U.S. Federal judge ruled that Dow, along with four other chemical companies, Monsanto, Thompson Hayward, Hercules Inc., and Diamond Shamrock, could be sued by Vietnam veterans. The veterans claim that they were injured by exposure to dioxin in the 2,4,5-T formulation "Agent Orange," and that this exposure resulted in a number of malformations among their children.[135] According to the veterans' attorney, the claim against the companies could be $40,000 million. The companies believe that the U.S. government should meet this bill and reimburse them for any damages incurred.[136] Dow argues that its contract was simply to supply herbicides to the government and that it had no say over the way in which the defoliants were used.[137]

If proof were needed that 2,4,5-T is safe, then Dow would claim to have shown beyond all reasonable doubt that the herbicide is indeed safe. Two mortality studies have been carried out on Dow workers who might have been contaminated with dioxin. In the first investigation, 204 workers known to have been involved in the manufacture of 2,4,5-T between 1950 and 1971 were traced.[130] Employment in the herbicide processing area ranged from a minimum of less than one year to a maximum of approximately ten years. None of the 204 men traced are reported to have had chloracne, a symptom of dioxin exposure. Deaths in this population are said by the authors to compare favorably with those for U.S. *white* males and also to compare well with the general mortality rate for Dow's Michigan complex.[130] Mortality in the 2,4,5-T process workers attributable to cancer and cardiovascular complaints was less than would be expected from a comparable control population. Only the six deaths from accidents and suicides were higher than the norm. However, the three deaths in separate motor vehicle accidents, the one death by fire (nonindustrial), and the two suicides, both more than ten years after working with 2,4,5-T, cannot be related to the handling of the herbicide.[130]

The second and more relevant study is an analysis of the mortality rates of a group of Dow workers known to have been exposed to dioxin. In an incident which occurred in 1964, 61 men employed on a production process for 2,4,5-trichlorophenol are thought to have been contaminated by the chemical.[138] In the case of 49 of the workers this exposure was certain—all developed chloracne. Since 1964, four of the 61 have died. The number of deaths expected for a comparable group is 7.8. Of the four men who died, one death was attributable to cardiovascular disease; 3.8 deaths would have been the norm. Three deaths were due to cancer, where only 1.6 fatalities could have been expected. None of the findings is statistically significant and little is revealed about the effect of dioxin exposure. Mortality trends for the incidence of cancer require much greater intervals between exposure to chemicals and the development of tumors than happened in this Dow survey. The group will require monitoring for some years to come to establish a more valid mortality trend.

Nevertheless, the findings in this mortality study have revealed some disturbing facts. Of the three deaths attributable to cancer in this group of Dow employees one was caused by a soft tissue sarcoma.[139] This type of tumor is a rare one, found in the main in fibrous tissue, fat, peripheral nerves, muscle, blood, synovium, and lymphatic tissue. For most tumors of this type the agent which caused the tumor is unknown. What is known about the Dow worker who died is that he had a fibrous tissue sarcoma and that he died 11 years after exposure to dioxin. Whether his death and his exposure to dioxin are related must remain speculation for the moment.

But death from a soft tissue sarcoma is not an isolated event for dioxin-exposed workers. At least two men who were exposed to the chemical at the Monsanto company have died from this type of tumor. In these two cases their deaths occurred 30 and 32 years after their first exposure to dioxin.[140] Dow, too, has had a second case with a soft tissue sarcoma. A worker employed at the company's Michigan complex who had his first contact with dioxin in 1951 has since developed a malignant fibrous histiocytoma. This brings to four the number of soft tissue sarcoma cases in dioxin-exposed workers. All of the men concerned had a history of smoking and it has been suggested that smokers exposed to dioxin may be at greater risk of developing soft tissue sarcoma.[140] Based on the current evidence (see Monsanto above) the incidence of soft tissue cancers in workers exposed to dioxin in U.S. chemical companies is 41 times that of the normal population, a significant risk factor.

Holmesburg Prison For those serving time in a penal institution the risk of exposure to dioxin would seem to be one of the least of their problems. Unless of course, the exposure was deliberate. For one group of men in Holmesburg Prison in Lewisburg, Pennsylvania, their exposure to dioxin was indeed deliberate. It seems that the Holmesburg inmates were invited to take part in one of the more unusual experiments involving dioxin.

Testing drugs on prison volunteers is no longer accepted practice in the United States. Following publicity given to cases where drug testing schemes were being abused, the practice was stopped. In the middle 1960s however, these constraints were not yet in operation and Dow was able to take advantage of the situation. The company was concerned about the 1964 outbreak of chloracne in its workforce on the trichlorophenol plant. Dioxin, Dow believed, was almost certainly the cause of the skin disease. But what concentration of dioxin was required to cause this? All Dow knew was that 0.5 μg of dioxin applied to the inside of a rabbit's ear caused mild chloracne. Application of 1–2 μg of the chemical caused a more pronounced effect, and 4–8 μg a severe one. The rabbit ear, as others had shown, was clearly sensitive to dioxin. But just how applicable were these results to humans? To get an answer, Dow approached Dr. Albert Kligman, Professor of Dermatology at the University of Pennsylvania, to inquire whether he would test dioxin for its ability to cause chloracne on prison volunteers. Kligman agreed. He would test dioxin on volunteers in Holmesburg Prison, a venue he was familiar with from previous drug testing schemes.

Aware that it was dealing with a chemical known to be extremely toxic, Dow drew up a careful protocol for testing dioxin. Only volunteer prisoners were tested, and each was asked to sign a consent form. Dr. V. K.

Rowe, the Dow scientist in charge of this testing program, informed the Environmental Protection Agency at its hearings on dioxin in November 1980 that this protocol for testing was "highly conservative."[141] The doses to be used were the same as those which produced a response in the rabbit. Rowe stressed that the highest dose to be tried was 8 μg. This dose could be given on a second occasion if no response was seen after the first application of the chemical to the skin. Providing this testing schedule was followed, no individual would be exposed to more than 16 μg of dioxin. Rowe also stressed that it was important to watch individuals for any adverse clinical symptoms—in particular, effects on the liver or kidney.

Sixty volunteers—six groups of ten men—participated in the program between late 1965 and early 1966. Concentrations of between 0.2 and 8 μg of dioxin were applied to either the forearm or mid-back region and the site covered for 24 hours with gauze. Two weeks later the practice was repeated, after which individuals were closely monitored for six weeks— that is, those volunteers who remained in the study were followed for this time. A few, however, left after the first application of dioxin and were not medically examined after this.[142]

None of the individuals developed chloracne. Fortunately none of them showed any other symptoms. From these results it was clear that apparently healthy men could tolerate 16 μg of dioxin without ill effect. The amount of dioxin which would cause harm in man thus remained unknown.

Kligman was disappointed that all this work was to no avail; Dow sympathized with him. To show its appreciation, the company agreed to fund further investigations by Kligman. However, before a subsequent protocol could be agreed on, it appears that Kligman, acting on his own initiative, experimented with a higher concentration of dioxin on a further ten volunteers. This time the amount used was 7,500 μg of dioxin applied to the back in an area one inch square. The new dose was almost one thousand times higher than the maximum stipulated by Rowe for the first study. Not surprisingly perhaps, eight of the ten volunteers now developed chloracne. The skin disease—which lasted 4–7 months—was the only apparent medical complication these men suffered.

Rowe was informed of these results in January, 1968. The work was unsolicited, and Rowe was quite clearly taken aback by it. Kligman had acted on his own to obtain this information—information which is of doubtful value. The two series of tests merely show that a dose of somewhere between 16 and 7,500 μg of dioxin is required to cause chloracne in man—hardly an exact figure for a threshold dose.

In volunteering this information to the EPA hearings, Dow was clearly taking a risk that it might be censured. Thankfully the company decided to go ahead and present the evidence. Rowe—appearing for Dow at the hearings—admitted that he had no further information on this group of prison volunteers. Nor did he have any information about their current whereabouts. It is perhaps a little strange that the company does not have this information. One can only surmise that Dow was alarmed at the way Kligman had conducted his second series of tests without informing the company and to prevent any repetition decided to end its relationship with him. As for the prisoners exposed to dioxin, one hopes that some effort will be made to trace them, if for no other reason than that they were all exposed to a carcinogen.

Monsanto—Newport, South Wales

The Monsanto Company has not had a great deal of luck with its processes for manufacturing chlorinated phenols. In August 1978 following discussions with the U.K. Health and Safety Executive concerning potential health hazards during the manufacture of pentachlorophenol (PCP) the company decided to stop production. Monsanto claimed that closure of the 3000-ton/year plant was already under consideration and that their discussions with the HSE merely hastened the process.[143] Twelve definite and eight suspected cases of chloracne had occurred at the plant a few years before closure (production at the Newport site began in 1950). Monsanto had already spent £500,000 on improvements to the plant before the HSE asked for a temporary suspension in order to assess the health of the plant workers.[143]

An assessment of the health of the Monsanto workers has been made by Monsanto and the HSE's Employment Medical Advisory Service. Forty men employed on the PCP—including pentachlorophenol and pentachlorophenate (Santobrite)—process for more than three months constituted the potential dioxin exposed group. (The chlorinated dioxins present in the pentachlorophenol process include hexa-, hepta, and octachlorodibenzodioxins. Hexachlorodibenzodioxins are known to cause chloracne.) Some 25 workers who may have had contact with the PCP process and 23 workers with no significant contact constituted two forms of control group.[144] Results of the tests show that in the dioxin-exposed group those with chloracne were at greater risk of developing ischaemic heart disease than any of the other groups. The workers with chloracne had higher blood cholesterol and blood triglyceride levels than their matched peers. Elevated blood cholesterol levels are considered to be a risk factor for heart disease, as is a reduction in the blood high-density lipoprotein (HDL) level. HDL values were more variable in the chloracne group than the controls.[144] All the workers at the plant were informed about the findings and individuals with abnormalities in the factors described above were counseled to reduce weight and fat consumption, stop smoking, and to see their family doctor.

Monsanto does not consider these results sufficiently striking to be of interest to the scientific or medical press. According to Dr. Alex Munn, Medical Director of Monsanto in Europe, the "findings are not sufficient to warrant publishing."[145] The HSE is not of the same opinion. The Executive considers that the Monsanto survey will yield useful "scientific knowledge about the consequence to workers of exposure to dioxin."[146] In view of its involvement in the investigation, the Executive intends to publish the results as part of a general review on dioxins.

Ministry of Agriculture Laboratories—United Kingdom

Exposure to dioxin need not occur in the confines of a factory nor in 2,4,5-T-sprayed forests as Dr. Donald Lee, a chemist employed by the U.K. Ministry of Agriculture, will attest. Lee, who worked at the Plant Pathology Laboratory at Harpenden, says that he and two colleagues were exposed to dioxin in 1970 in the course of an investigation of the toxicity of 2,4,5-T.[147] The reason for Lee's interest in 2,4,5-T was a request from Forestry Commission workers for details about the dioxin content of the herbicide. Reports from Vietnam about the herbicide Agent Orange had worried workers and Lee was asked to investigate.

To do his analyses Lee required standard samples of dioxin. As these were not available from commercial suppliers, he decided to make his own. Aware of the toxicity of the material he was preparing, Lee and a colleague took what they considered to be adequate safety precautions. Different methods of preparation were used by the two men. Lee chose to heat potassium trichlorophenate in a closed system taking care to avoid inhalation and skin contact with the reaction mixture.[148] His colleague heated trichlorophenol in an alkaline solution in the presence of a catalyst. Five to eight weeks after their respective episodes in the laboratory both men noticed changes in their skin condition. Lee's skin became excessively oily, while his colleague's displayed all the signs of chloracne—blackheads and cysts appeared on his forehead, neck, behind the ears, and to a lesser extent on his arms and trunk. A few weeks later Lee also developed chloracne; the skin disease appeared in two waves about a week apart, coinciding with the two occasions—eight weeks previously—when he was involved in preparing the pure dioxin standard. In addition to the skin disease, which gave off the characteristic rancid odor, he experienced pains in his abdomen and flatulence when consuming breakfast foods. The condition was most marked with foods containing oats. Lee's mental and muscle coordination were impaired and he noted abnormal hair growth on the shoulders, upper part of the back, around the nipples and on the eyebrows. To make matters worse his concentration was diminished and he became easily irritable and prone to episodes of uncharacteristic anger.[147]

Lee, the senior scientist in the laboratory, experienced symptoms far more serious than his colleague, and much more severe than a second laboratory technician who was contaminated when handling diluted samples of the dioxin standard. He, too, experienced flatulence, had an oily skin, but had no chloracne, and developed longer, coarser hair growth on the shoulders, upper back, and on the eyebrows. Lee was particularly concerned by the change in his blood cholesterol. It rose to a value of 305 mg/100 ml [7.9 mmol/liter (millimoles per liter). The normal age-related range for 30-39-year-olds is about 3.6-7.0 mmol/liter. High blood cholesterol is a recognized risk factor for heart disease.] Both of his colleagues had similar high values. In commenting on these figures Lee is quoted as saying "we are all quite young (Lee was 38 at the time, the others were 28 and 30, respectively). None of us is obese. We are all fit, we are not exactly sports maniacs but we play squash, and I spend all my spare time digging in the garden. I am fairly lean for my age, yet I had these huge cholesterol levels. Why?"[148]

The question was rhetorical; the answer was almost certainly dioxin.

Hooker, Thompson Hayward, Bayer, and Chemie Linz

In the case of four other industrial accidents, details are only very scanty. Hooker Chemicals at Niagara Falls, New York, is known to have experienced an accident in its 2,4,5-trichlorophenol plant in 1956.[149] Thompson Hayward experienced similar problems in 1959. But when officials from the U.S. National Institute for Occupational Safety and Health asked Thompson Hayward about the incident in the summer of 1979, a company spokesman denied that it had ever occurred.[150] Later, however, more knowledgeable employees confirmed that there had been an accident 20 years previously.[151]

Poor hygiene is reputed to be responsible for workers suffering from dioxin exposure in 1970 at the Vedinger site of the West German company Bayer,[149] and between 1972 and 1973 at the Austrian firm of Chemie-Werke at Linz.[149] First reports suggested that 50 workers were affected at the Austrian plant, but informed sources put the total nearer to 100.

Conclusion

Industrial incidents involving the release of dioxin have not been unusual, as this chapter has shown. Although poor hygiene, rather than actual explosions, has been the major cause of worker exposure to dioxin, both situations illustrate that the 2,4,5-trichlorophenol process is costly in human terms. Two thousand men have developed a most disfiguring skin condition, chloracne, as a result of their involvement with the chemical process. What the future holds for them, no one can be absolutely sure. Dioxin will cause blood cholesterol levels to rise and thus increase the workers' risk of developing ischaemic heart disease. The chemical causes cancer in animals and must, for that reason, be a suspected carcinogen in humans. Whether dioxin-exposed workers have a higher incidence of heart attacks and cancer is not clear from the evidence available so far. There is a suggestion, however, that at least one group of dioxin-exposed workers in Holland does have a higher incidence of deaths from heart attacks. As for cancer, the evidence suggests that men exposed to dioxin have a very high incidence of soft tissue cancer. Where deaths directly attributable to explosions in trichlorophenol reactor vessels are concerned, only one has been reported. Eric Burrows, the supervisory chemist at the Coalite plant in the United Kingdom, was killed by masonry dislodged in the explosion which destroyed the reactor at the Bolsover site in 1968.

Disturbing as these medical problems are they have been made worse by the attitude of many of the companies involved. Only a handful have taken the course of making freely available all the details of their accidents for the benefit of other manufacturers. Had the Seveso accident not occurred, few details would have emerged about the various incidents. Indeed, in the aftermath of the Seveso explosion, little was in fact known about accidents involving dioxin, a situation which made life exceedingly difficult for officials in Italy with responsibility for the health of the Seveso residents.

But the most worrisome aspect of all must be the deliberate withholding of information. At Coalite and Chemical Products Limited, the management took the decision not to publish details of three medical reports on its dioxin-exposed workforce. It was only after details about the findings of the medical reports and Coalite's decision not to publish them were reported in the scientific journal *Nature*[152] that pressure was put on the company to divulge the information by the U.K. Health and Safety Executive. The three reports were given to the Executive some weeks later. Other companies are not blameless. Monsanto in West Virginia at one time was not particularly forthcoming about the health problems in its workforce. It took the Seveso accident and the pressure of other companies and health officials to persuade the company to adopt a more open policy.

Openness, or at least a willingness to show medical records to government health officials, ought now to be the rule for all companies, in acceptance of responsibility for the welfare of their workforce. A pooling of information will be needed to unravel the long-term health risks faced by dioxin-exposed workers. This could be achieved by more enlightened attitudes on the part of company management. Sadly, it seems that in many cases it will still require pressure from workers' trade unions and government health and safety organizations to force companies to divulge information.

References

1. Hay, A. W. M. Tetrachloradibenzo-*p*-dioxin release at Seveso, *Disasters* **1**, 284–308 (1977).
2. Young, A. L., Calcagni, J. A., Thalken, C. E., and Tremplay, J. W. The toxicology, environmental fate, and human risk of herbicide orange and its associated dioxin, Report OEHL TR-78-92, USAF Occupational and Environmental Health Laboratory, Aerospace Medical Division (AFSC), Brooks Air Force Base, Texas 78735 (1978).
3. Holmstedt, B. Prolegomena to Seveso, *Archives of Toxicology* **44**, 211–230 (1980).
4. Ashe, W. F., and Suskind, R. R. Report—Patients from Monsanto Chemical Company, Nitro, West Virginia, Internal Monsanto report (5 December 1949).
5. Ashe, W. F., and Suskind, R. R. Progress Report—Patients from Monsanto Chemical Company, Nitro West Virginia (April 1950).
6. Suskind, R. R. A review of occupational exposures to dibenzo-*p*-dioxins, presentation to a conference on dibenzodioxins/dibenzofurans, Rougemont, North Carolina (18 November 1976).
7. Zack, J. A., and Suskind, R. R. The mortality experience of workers exposed to tetrachloradibenzo-dioxin in a trichlorophenol process accident, *Journal of Occupational Medicine* **22**, 11–14 (1980).
8. Baader, E. W., and Bauer, H. J. Industrial intoxication due to pentachlorophenol, *Industrial Medicine and Surgery* **20**, 286–290 (1951).
9. Bauer, H., Schultz, K. H., and Spielgalberg, U. Occupational intoxication in the manufacture of chlorophenol compounds, *Archiv fuer Gewerbepathologie und Gewarbehygiene* **18**, 538–555 (1961).
10. Hay, A. W. M. Exposure to TCDD: The health risks, in *Chlorinated Dioxins and Related Compounds* by O. Hutzinger (ed.) Pergamon Press, Oxford and New York (1982), pp. 589–600.
11. Goldmann, P. J. Schwerste akute Chlorakne durch Trichlorphenol—Zersetzungsprodukte, *Arbeitsmedizin Sozialmedizin Arbeitshygiene* **7**, 12–18 (1972).
12. Goldmann, P. J. Schwerste akute Chlorakne, eine Massenintoxikation durch 2,3,6,7-Tetrachlor-dibenzodioxin, *Hautarzt* **24**, 149–152 (1973).
13. Thiess, A. M., and Frentzel-Beyme, R. Mortality study of persons exposed to dioxin after an accident which occurred in the BASF on 13.11.1953, Vth International Medichem Congress, San Francisco (5 September 1977) (follow-up study of workers exposed in 1953).
14. IARC: Joint NIEHS/IARC Working Group Report, Long-term hazards of polychlorinated dibenzodioxins and polychlorinated dibenzofurans, International Agency for Research on Cancer, Lyon, France (1978).
15. Dugois, P., Marechal, J., and Colomb, L. Chloracne caused by 2,4,5-trichlorophenol, *Archives des Maladies Professionnelles Hygiene et Toxicologie Industrielles* **19**, 626–627 (1958).
16. Kimmig, J., and Schulz, K. H. Occupational acne (so-called chloracne) due to chlorinated aromatic cyclic ethers, *Dermatologica* **115**, 540–546 (1957).
17. Kimmig, J., and Schulz, K. H. Chlorierte aromatische Zyklische Ather als Ursache der sogenannten Chlorakne, *Naturwissenschaften* **44**, 337–338 (1957) (the important summary by Schulz of the properties of dioxin (TCDD)).

18. Kraus, Von L., and Brassow, H. Katamnestischer Beitrag zu sog. Chlorakneerkrankungen aus dem Jahre 1954/55, *Arbeitsmedizin Socialmedizin Praeventivmedizin* **13**, 19-21 (1978).
19. Bleiburg, J., Wallen, M., Brodkin, R., and Appelbaum, I. L. Industrially acquired porphyria, *Arch. Dermatol.* **89**, 793-797 (1964).
20. Poland, A. P., Smith, D., Metter, G., and Possick, P. A health survey of workers in a 2,4-D and 2,4,5-T plant, *Archives of Environmental Health* **22**, 316-327 (1971).
21. Hofman, M. F., and Meneghini, C. L. A proposito delle follicolosi da idrocarburi clorosostituito (acne chlorica), *Giornale Italiano di Dermatologia* **103**, 427-450 (1962).
22. Dalderup, L. M. Safety measures for taking down buildings contaminated with toxic material, I. *T. Soc. Geneesk.* **52**, 582-588 (1974).
23. Dalderup, L. M. Safety measures for taking down buildings contaminated with toxic material, II. *T. Soc. Geneesk.* **52**, 616-623 (1974).
24. Vos, J. G., Workshop on Dibenzodioxins and dibenzofurans, paper presented at the International Agency for Research on Cancer (reported in Reference 14).
25. Telegina, K. A., and Bikbulatova, L. I. Afflication of the follicular apparatus of the skin in workers occupied in the production of butyl ether of 2,4,5-trichlorophenoxyacetic acid, *Vestnik Dermatologii i Venerologii* **44**, 35-39 (1970).
26. Firestone, D. The 2,3,7,8-tetrachlorodibenzo-para-dioxin problem. A review, in Chlorinated phenoxy acids and their dioxins: Mode of action, Health risks and environmental effects, *Ecol. Bull.* (Stockh.) **27** (39-52 (1978).
27. Jirasek, L., Kalensky, J., and Kubec, K. Acne chlorina and porphyria cutanea tarda during the manufacture of herbicides, *Ceskoslovenska Dermatologie* **48**, 306-317 (1973).
28. Jirasek, L., Kalensy, J., Kubeck, K., Pazderova, J., and Lukas, E. Acne chlorina porphyria cutanea tarda and other manifestations of general intoxication during the manufacture of herbicides. II *Ceskoslovenska Dermatologie* **49**, 145-157 (1974).
29. Jirasek, L., Kalensky, J., Kubeck, K., Pazderova, J., and Lukas, E. Chlorakne, porphyria cutanea tarda und andere intoxikationen durch herbizid., *Hautarzt* **27**, 328-323 (1976).
30. May, G. Chloracne from the accidental production of tetrachloradibenzodioxin. Properties of TCDD, *British Journal of Industrial Medicine* **30**, 276-283 (1973).
31. Oliver, R. M. Toxic effects of 2,3,7,8-tetrachlorodibenzo-1,4-dioxin in laboratory workers, *British Journal of Industrial Medicine* **32**, 49-53 (1975).
32. Mivra, H., Omori, A., Shibue, M. The effect of chlorophenols on the excretion of porphyrins in urine, *Japanese Journal of Industrial Health* **16**, 575-577 (1974).
33. Zelikov, A. K., and Danilov, L. N. Occupational dermatoses (acnes) in workers engaged in production of 2,4,5-trichlorophenol, *Sovetskaya Meditsina* **7**, 145-146 (1974).
34. Hay, A. Toxic cloud over Seveso, *Nature (London)* **262**, 636-638 (1976).
35. Hay, A. Seveso: the aftermath, *Nature (London)* **263**, 538-540 (1976).
36. Reggiani, G. Toxic effects of TCDD in man, presented to a NATO Workshop on Ecotoxicology, Guildford, England (August 1977).
37. Minutes of Chloracne meeting on 7 November 1979, Health and Safety Executive.
38. Theiss, A. M., and Frentzel-Beyme, R. Mortality study of persons exposed to dioxin after an accident which occurred in the BASF on 13 November 1953, 5th International Medichem Congress, San Francisco (5-9 September 1977).

Monsanto

39. Zack, J. A., and Suskind, R. R. The mortality experience of workers exposed to tetrachlorodibenzodioxin in a trichlorophenol process accident, *Journal of Occupational Medicine* **22**, 11-14 (1980).

40. Suskind, R. R. Personal letter, 29 February 1980.
41. Ashe, W. F., and Suskind, R. R. Report—Patients from Monsanto Chemical Company, Nitro, West Virginia (5 December 1949).
42. Oliver, R. M. Toxic effects of 2,3,7,8-tetrachlorodibenzo-1,4-dioxin in laboratory workers, *British Journal of Industrial Medicine* **32**, 49–53 (1975).
43. Walker, A. E., and Martin, J. V. Lipid profiles in dioxin-exposed workers, *Lancet* **I**, 446 (1979).
44. Ashe, W. F., and Suskind, R. R. Progress report—Patients from Monsanto Chemical Company, Nitro, West Virginia (April 1959).
45. Crow, K. D. Unpublished data (1979).
46. Honchar, P. A., and Halperin, W. E. 2,4,5-T, trichlorophenol and soft tissue sarcoma, *Lancet* **I**, 268–269 (1981).
47. Cook, R. R. Dioxin, chloracne and soft tissue sarcoma. *Lancet* **I**, 618–619 (1981).
48. Zack, J., and Suskind, R. R. Details given in pr newswire, St. Louis (9 October 1980) (to be published).

BASF

49. BASF Annual Report (1977).
50. Borkin, J. *The Crime and Punishment of IG Farben*, The Free Press, New York and London (1978).
51. Theiss, A. N., and Frentzel-Beybe, R. Mortality Study of Persons Exposed to Dioxin after an Accident which Occurred in the BASF on 13 November 1953, Vth International Medichem Congress, San Francisco (5–9 September 1977).
52. Goldmann, Von, P. J. Schwerste skute durch Trichlorophenolzersetzungsprodukte, *Arbeitsmedizin Sozialmedizin Arbeitshygiene* **7**, 12–18 (1972).
53. Goldmann, P. J. Schweste skute chloracne, eiene Massenintoxikation durch 2,3,6,7-tetrachlorodibenzodioxin, *Hautarzt* **24**, 149–152 (1973).
54. Hay, A. Dioxin meeting recommends cancer study, *Nature (London)* **271**, 202 (1978).
55. International Agency for Research and Cancer, Long term hazards of polychlorinated dibenzodioxins and polychlorinated dibenzofurans, IARC Internal Technical Report No. 78/001 (June 1978).
56. Theiss, A. M., Frentzel-Beyme, R., and Link, R. Mortality study of persons exposed to dioxin in a trichlorophenol-process accident which occurred in the BASF AG on November 17, 1953 (in press).
57. Rohrborn, G., Goldmann, P., and Fleig, I. Chromosomenanalyse bei Mitarbeitern mit Exposition gegenuber 2,3,7,8-tetrachlorodibenzodioxin (TCDD)—Umfallgeschehen am 17 November 1953 (unpublished).
58. Huff, J. E. Personal letter, 2 October 1978.

Philips-Duphar

59. Dalderup, L. M. Safety measures for taking down buildlings contaminated with toxic material I, *T. Soc. Geneesk* **52**, 582–585 (in English).
60. Dalderup, L. M. Safety measures for taking down buildings contaminated with toxic material II, *T. Soc. Geneesk* **52**, 616–623 (in English).
61. Masterman, S. *The Times*, 9 August 1976.
62. Hay, A. Toxic cloud over Seveso, *Nature (London)* **262**, 636–638 (1976).
63. Vos, J. G., Sterringa, Tj., Zellenrath, D., Docter, H. J., and Dalderup, L. M. TCDD accident at a chemical factory in the Netherlands, presented at a joint National Institute of

Environmental Health Sciences/International Agency for Research on Cancer Working Group on the Long Term Hazards of Chlorinated Dibenzodioxins and Dibenzofurans, IARC, Lyon (10-11 January 1978).
64. Schuuring, C. Toxic waste: Dutch dumps, *Nature (London)* **289**, 340 (1981).
65. Barnaby, W. Personal communication (April 1981).

Coalite and Chemical Products Limited

66. Anonymous, *The Sheffield Star*, 24 April 1968.
67. Coalite Group, Report and Accounts 1977/1978.
68. Milnes, M. H. Formation of 2,3,7,8-tetrachlorodibenzodioxin by thermal decomposition of sodium trichlorophenate, *Nature (London)* **232**, 395-396 (1971).
69. May, G. Chloracne from the accidental production of tetrachlorodibenzodioxin, *British Journal of Industrial Medicine* **30**, 276-283 (1973).
70. Anonymous, *The Sheffield Morning Telegraph*, 17 May 1968.
71. Adams, E. M., Irish, D. D., Spencer, H. C., and Rowe, U. K. The response of rabbit skin to compounds reported to have caused acne form dermatitis, *Industrial Medicine*, Industrial Hygiene Section 2, 1-4 (1941).
72. Anonymous, *The Sheffield Star*, 2 August 1976.
73. Anonymous, *The Sheffield Star*, 30 July 1976.
74. Needham, C. Personal interview, 22 July 1977.
75. *The Times, The Guardian, The Daily Telegraph*, 13 August 1976.
76. Anonymous, *The Sheffield Morning Telegraph*, 15 October 1976.
77. Anonymous, *The Sheffield Morning Telegraph*, 7 August 1976.
78. Anonymous, *The Sheffield Star*, 23 October 1976.
79. Young, A. L., Tahlken, C. E., Arnold, E. L., Cupello, J. M., and Cockerman, L. G. Fate of 2,3,7,8-tetrachlorodibenzo-*p*-dioxin (TCDD) in the environment: summary and decontamination recommendations, USAFA-TR-76-18, Department of Chemistry and Biological Sciences, USAF Academy, Colorado 80840 (1976).
80. Di Domenico, A., Silano, V., Viviano, G., and Zapponi, G. Accidental Release of 2,3,7,8-tetrachlorodibenzo-*p*-dioxin (TCDD) at Seveso (Italy): V. Environmental Persistence (Half Life) of TCDD in Soil, Instituto Superiore di Sanita (26 January 1980).
81. Caufield, C., and Pearce, F. An overburden of toxic waste, *New Scientist* **90**, 344-347 (1981).
82. Needham, C. Television interview, 7 May 1981.
83. Anonymous, *The Sheffield Star*, 11 August 1976.
84. Anonymous, *The Sheffield Morning Telegraph*, 8 October 1976.
85. Anonymous, *The Sheffield Star*, 23 December 1976.
86. Anonymous, *The Sheffield Star*, 31 December 1976.
87. Anonymous, *The Sheffield Morning Telegraph*, 10 September 1976.
88. Marshall, R. P. Personal interview, October 1978.
89. Ashby, J., Styles, J. A., Elliot, B., and Hay, A. W. M. (unpublished).
90. Kociba, R. J., Keyes, D. G., Beyer, J. E., Carreon, R. M., Wade, C. E., Dittenber, D. A., Kalnins, R. P., Frauson, L. E., Park, C. N., Barnard, S. D., Hummel, R. A., and Humiston, C. G. Results of a two year chronic toxicity and oncogenicity study of 2,3,7,8-tetrachlorodibenzo-*p*-dioxin (TCDD) in rats, *Toxicology and Applied Pharmacology* **46**, 279-303 (1978).
91. Blank, C. E. Bolsover Project 1977-1978, Genetic damage (Report to Coalite).
92. Ward, A. M. Investigation of the immune capability of workers previously exposed to 2,3,7,8-tetrachlorodibenzo-*p*-dioxin (TCDD) (Report to Coalite).

93. Martin, J. V. Report of a biochemical study carried out on workers at Coalite and Chemical Products Limited, Chesterfield, Derbyshire (Report to Coalite).
94. Walker, A. E., and Martin, J. V. Lipid profiles in dioxin-exposed workers, *Lancet* **I**, 446 (1979).
95. Hay, A. Chemical company suppresses dioxin report, *Nature (London)* **284**, 2 (1980).
96. Anonymous. A study of workers who have been, or were liable to have been exposed to dioxin, carried out, on behalf of Coalite and Chemical Products Limited, at the plant of the manufacturing subsidiary company, Coalite Oils and Chemicals Limited (November, 1978).
97. Ratcliffe, R. *Sunday Times*, 9 March 1980.
98. Ratcliffe, R. *Sunday Times*, 16 March 1980.
99. Anonymous, *Yorkshire Post*, 20 March 1980.
100. Hay, A. Coalite health survey talks, *Nature (London)* **285**, 4 (1980).
101. New release, Health and Safety Executive (1 May 1980).
102. Hay, A. Dioxin hazards: secrecy at Coalite, *Nature (London)*, **290**, 729 (1981).
103. Bain, S. *The Sheffield Star* (27 June 1980).
104. Report on Coalite Chemicals, Health and Safety Office, ASTMS (May 1980).

Soviet Union

105. Telegina, K. A., and Bikbulatova, L. I. Effects on the follicular apparatus of the skin of workers occupied in the production of the butyl ester of 2,4,5-trichlorophenoxyacetic acid (English translation), *Vestnik Dermatologii i Venerologii* **44**, 35–39 (March 1970).
106. Crow, K. D. Chloracne: the chemical disease, *New Scientist*, 78–80 (13 April 1978).

Missouri

107. Kimbrough, R. D., Carter, C. D., Liddle, J. A., Cline, R. E., and Philips, P. E. Epidemiology and pathology of a tetrachlorodibenzodioxin poisoning episode, *Archives of Environmental Health* **32**, 77–86 (1977).
108. Commoner, B., and Scott, R. E. Accidental contamination of soil with dioxin at Missouri: effects and countermeasures, a contribution to the Dioxin Information Project, Scientists' Institute for Public Information, New York (29 September 1976).
109. Donnel, H. D., and Philips, P. E. Draft report to the Director of the Missouri State Division of Health from the section of epidemiology (11 September 1974).
110. Beale, M. A., Shearer, W. T., Karl, M. M., and Robson, A. M. Long term effects of dioxin exposure, *Lancet* **I**, 748 (1977).
111. Lobes, L. A., Koehler, R. F., *et al.* Administrative report to the Director of the Center for Disease Control (CDC) on Toxic illness, Lincoln County, Missouri, CDC Number EPI-72-13-2, U.S. Public Health Service, CDC, Atlanta, Georgia (14 August 1972).
112. Report to Director, Centers for Disease Control, from Cancer and Birth Defects Division, Bureau of Epidemiology, and Clinical Chemistry Division, Bureau of Laboratories, HEW, Public Health Service—CDC—Atlanta, EPI-75-17-2 (31 March 1975).
113. Letter to Senator Philip A. Hart, Chairman, Sub-committee on the Environment, Committee on Commerce, United States Senate from David J. Spencer, MD, Assistant Surgeon General, Director, Department of Health, Education and Welfare (15 September 1974).
114. Memorandum to Chief, Bacterial Diseases Branch from Renate D. Kimbrough, HEW, Public Health Service, Centers for Disease Control, Atlanta, Georgia (30 July 1974).

115. Letter to Mr. Paul E. des Rosiers, Staff Engineer, Office of Energy, Minerals and Industry Environmental Protection Agency, from Renate D. Kimbrough (12 June 1975).
116. Kimbrough, R. D. Personal letter (11 September 1979).
117. EPA official, personal interview (3 March 1980).
118. des Rosiers, P. E. Personal interview (11 May 1980).

Diamond Alkali

119. Poland, A. M., Smith, D., Mether, G., and Possick, P. A health survey of workers in a 2,4-D and 2,4,5-T plant, *Archives of Environmental Health* **22**, 316–327 (1971).
120. Bleiberg, J., Wallen, M., Brodkin, R., and Appelbaum, L. Industrially acquired porphyria, *Archives of Dermatology* **89**, 793–797 (1964).

Spolana

121. Jirasek, L., Kalensky, J., and Kubeck, K. Acne chlorine and porphyria cutanea tarda during manufacture of herbicides, *Ceskoslovenska Dermatologie* **48**, 306–317 (1973).
122. Jirasek, L., Kalensky, J., Kubeck, K., Pazderova, J., and Lukas, E. Acne chlorine, porphyria cutanea tarda and other manifestations of general intoxication during the manufacture of herbicides, *Ceskoslovenska Dermatologie* **49**, 145–147 (1974).
123. Jirasek, L. Presentation to Joint National Institute of Environmental Health Sciences/International Agency for Research on Cancer working group, on Long Term Hazards of Chlorinated Dibenzodioxins/Chlorinated Dibenzofurans, International Agency for Research on Cancer, Lyons (10–11 January 1978).
124. Pazderova-Vejlupkova, J., Nemcova, M., Pickova, J., Jirasek, L., and Lukas, E. The development and progress of chronic intoxication by tetrachlorodibenzo-*p*-dioxin in men, *Archives of Environmental Health* **36**, 5–11 (1981).
125. Spontaneous and induced abortion, Report of a WHO Scientific Group, World Health Organization Technical Report Series, No. 461, Geneva (1970).

Boehringer

126. Kraus, Von L., and Brassau, H. Katamnestischer Beitrag zu sog Chlorakneer Krankungen aus dem Jahre 1954/55, *Arbeitsmedizin Socialmedizin Praventivmedizin* **13**, 19–21 (1978).

Rhone-Poulenc

127. Dugois, P., Marechal, J., and Colomb, L. Chloracne caused by 2,4,5-trichlorophenol, *Archives des Maladies Professionelles Hygiene et Toxicologie Industrielles* **19**, 626–627 (1958).
128. Dugois, P., and Colomb, L. Acne chlorique au 2,4,5-trichlorophenol, *Lyon Medical* **88**, 446–447 (1956).
129. Dugois, P., and Colomb, L. Remarques sur l'acne chlorique (a propos d'une eclosion de cas provoques par la preparations du 2,4,5-trichlorophenol), *Le Journal de Medicine de Lyon* **38**, 899–903 (1957).

Dow Chemical

130. Ott, M. G., Holder, B. B., and Olson, R. D. A mortality analysis of employees engaged in the manufacture of 2,4,5-trichlorophenoxyacetic acid, *Journal of Occupational Medicine* **22**, 47–50 (1980).
131. Cookson, C. Emergency ban on 2,4,5-T herbicide in U.S., *Nature (London)* **278**, 108–109 (1979).
132. Hay, A. Critics challenge data that led to 2,4,5-T ban, *Nature (London)* **279**, 3 (1979).
133. Editorial: Dioxin and 2,4,5-T: What are the Risks? *Nature (London)* **284**, 111 (1980).
134. Patton, D. E., Lee, K. M., Roberts, P. A. Bozof, R. P., Backstram, T. D., and Morrison, S. E. Respondents' prehearing brief on the risks associated with the registered uses of 2,4,5-T and Silvex, U.S. Environmental Protection Agency before the Administrator FIFRA Docket Nos. 415 *et al.*
135. Dioxin traces found in soldiers exposed to defoliant, *Nature (London)* **282**, 772 (1979).
136. Defoliant companies can be sued, *Nature (London)* **282**, 434 (1979).
137. Kavanagh, K. Office of Victor Yannacome, personal communication to the author (2 June 1980).
138. Cook, R. R., Townsend, J. C., Ott, M. G., and Silverstein, L. G. Mortality experience of employees exposed to 2,3,7,8-tetrachlorodibenzo-*p*-dioxin (TCDD), *Journal of Occupational Medicine* **22**, 530–532 (1980).
139. Honchar, P. A., and Halperin, W. E. 2,4,5-T, trichlorophenol and soft tissue sarcomas, *Lancet* **I**, 268–269 (1981).
140. Cook, R. R. Dioxin, chloracne and soft tissue sarcoma. *Lancet* **I**, 618–619 (1981).
141. Rowe, V. K. Direct Testimony of Dr. V. K. Rowe (Exhibit 865) before the Environmental Protection Agency of the United States of America. 13 November 1980.
142. Transcript of U.S. Environmental Protection Agency Hearings in Re: The Dow Chemical Company *et al.* p. 17087. November 1980.

Monsanto—Newport

143. Monsanto stops PCP plant at Newport, *European Chemical News*, 1 September 1978, p. 7.
144. Notes of an informal meeting to discuss chloracne, Health and Safety Executive, 7 November 1979.
145. Dr. Alex Munn, personal communication, January 1980.
146. Workers given EMAS assessment of survey of dioxin exposure at Coalite plant, Health and Safety Executive press release (1 May 1980).

Ministry of Agriculture, U.K.

147. Oliver, R. M. Toxic effects of 2,3,7,8-tetrachlorodibenzo 1,4-dioxin in laboratory workers, *British Journal of Industrial Medicine* **32**, 49–53 (1975).
148. Jacobs, P. *The Times*, 13 October 1976.

Hooker, Thompson Hayward, Bayer, and Chemie-Linz

149. Hay, A. W. M. Tetrachlorodibenzo-*p*-dioxin release at Seveso. *Disasters* **1**, 289–308 (1977).
150. Halperin, W. Personal letter, 23 July 1979.
151. Halperin, W. Personal letter, 14 September 1979.
152. Hay, A. Chemical company suppresses dioxin report, *Nature (London)*, **284**, 2 (1980).

7

Vietnam and 2,4,5-T

"Saddle up, Cowboys!" With this command from Lieutenant Colonel Jack Longhorne, lead pilot in a formation of five C-123 "Providers," the "Ranch Hands" of the 12th Special Operations Squadron prepared to rein into action. With the grace of hawks, the seemingly lumbering C-123's plummeted 3500 feet in less than a minute to level off just above the jungle canopy which shrouds the enemy-infested southernmost portion of the Mekong Delta. The spray jets were opened and defoliant spewed forth in another operation to dispense "herbicides, in use since 1962, (which) are non-toxic, non-corrosive and not harmful to human and animal life."[1]

This pen portrait, in a newsletter for United States servicemen in Vietnam dated January 1969, describes just one of the herbicide spraying missions, of which there were some 20,000 flown over the country. The aircrews in Operation Ranch Hand, the code name for the herbicide program, were instrumental in the application of some 19 million gallons of herbicide in Vietnam between 1962 and 1971.[2] The program was not a modest one. One study[3] claims 8.5% of the South Vietnamese countryside was sprayed one or more times, two others put the figure slightly higher at 10%.[4,5] According to the U.S. National Academy of Sciences 10.3% of Vietnam's inland forests, 36.1% of mangrove forests, 3% of cultivated land, and 5% of "other" land was affected by the spraying program.[3]

The U.S. Air Force claims that the herbicides were not harmful to humans and its insistence that missions were only flown "away from heavily populated areas"[5] were the constant rejoinders used to deflect any criticisms made about the effect of Operation Ranch Hand. But this complacency was rudely shattered with the publication in 1969 of a report claiming that one of the herbicides in use, 2,4,5-T (2,4,5-trichlorophenoxyacetic acid) had teratogenic (fetus-deforming) properties in rodents.[6] The report sounded the death knell for the herbicide spraying program. In a matter of weeks, President Nixon's science adviser, Dr. Lee Dubridge, announced restrictions on the use of 2,4,5-T; on 29 October 1969 it was stated that the U.S. Government would be taking action to restrict the use of the herbicide in civilian and military spraying programs.[7]

Thus, almost 30 years after research on the phenoxy herbicides was begun—a time during which the use of 2,4,5-T had increased exponentially—the military program which had done so much to publicize 2,4,5-T had become a millstone around the neck of the herbicide's manufacturers. The furor which developed following the publication of the teratogenicity study by the Bionetics Research Laboratory was due, in the main, to the fact that herbicide had been sprayed over inhabited areas, affecting some 3% of the population according to a Department of Defense background briefing paper.[5] The Bionetic's study was the first serious hitch for the herbicide since it was developed.

Investigations into the military use of herbicides began in the United States in late 1941. Prominent American scientists convinced Henry Stimson, who was then Secretary of War, of the potential dangers of biological warfare.[8] Stimson in turn approached the U.S. National Academy of Sciences and National Research Council for help and advice. In response the Academy formed a committee to investigate biological and chemical warfare—the so-called "ABC Committee." The committee acted swiftly, producing a report in February 1942 which led to the formation of the War Research Service.

Two scientists on the ABC Committee—Dr. E. J. Kraus, Head of the Botany Department at the University of Chicago, and one of his doctoral students, John Mitchell, working at the U.S. Department of Agriculture's Plant Industry Station at Beltsville, Maryland—played a prominent role in the investigations of the potential uses of phenoxy herbicides. Kraus noted the toxic properties of the two phenoxy herbicides 2,4,5-T and 2,4-D (2,4-dichlorophenoxy acetic acid) and suggested to the War Research Service that these chemicals might be used for destroying crops—enemy crops. He assumed that this would be of interest to the military.[8]

The control of vegetation by herbicide use was not even then a new practice. According to the NAS report on *The Effects of Herbicides in South Vietnam*,[2] the first recorded use of a chemical for selective weedkilling was in 1896. A French viticulturist observed that wild mustard was selectively killed by copper sulfate, at that time an ingredient of a fungicide "Bordeaux mixture." The next major step was the identification of the plant growth hormone auxin as 3-indole acetic acid. In high concentrations the growth hormones are toxic to plants; this observation was the basis for the development of synthetic acids suitable for use as weedkillers. 2,4,5-T and the closely related 2,4-D are two which have proved to be particularly effective for broad-leaved plants (but relatively innocuous where cereals and grasses are concerned).

Kraus' suggestions regarding the development of herbicidal chemicals were immediately taken up by the United States Army. Funds were provided for Kraus and Mitchell to investigate the matter. Meanwhile, an army research team at Fort Detrick was directed to intensify its research program; some 1100 substances were screened there. Of these, two chemicals, 2,4,5-T and 2,4-D, were shown to have outstanding herbicidal properties both in laboratory tests, and in field trials in the Florida Everglades.

But neither herbicide was used in World War II. One report claims that the war ended before the herbicides could be used.[8] But, there were other considerations too. Clearing vegetation on the islands in the Pacific had always been a problem for the U.S. Army. High explosives were the usual means employed but the Army was interested in subtler methods. By June 1945 scientists in the Army were prepared to recommend chemical defoliants as a suitable replacement. But the plans were dropped according to Dr. Charles Minarik of the Crops Division at Fort Detrick, in case the U.S. "would be accused of conducting poison gas warfare."[9]

In spite of the inhibitions it seems that the U.S. Government did have stocks of herbicide which it was preparing to use against the Japanese. A paper prepared for the British Joint Technical Warfare Committee in November 1945, and recently released, states that the U.S. stocks were considered sufficient "to destroy some 30% of the total rice crop (of Japan)."[10] The chemicals to be used included 2,4-D and 2,4,5-T. Significantly the document points out that the legal position with regard to the herbicides was clear; their use was not prohibited and there was no agreement on how, or when they could be used. The only restraining influences were moral ones. However, the document prepared for the Joint Chiefs of Staff had an eye to the future and it noted that the herbicides could be used "as a form of sanction against a recalcitrant nation which would be more speedy than blockade and less repugnant than the atomic bomb."[10]

Similar sentiments were probably expressed in U.S. military circles when the Korean War started in 1950. That there might be a need for vegetation control agents by the forces in the field was a sufficient excuse for the U.S. Air Force and Navy to increase their research effort to develop better systems for the aerial spraying of herbicides. By 1951 the favored herbicides were still the n-butyl esters of 2,4,5-T and 2,4-D. Delivery systems for these herbicides were ready for use in 1953, but in the event were never tested in the conflict itself. When the war ended in 1953 the delivery systems were placed in cold storage and the herbicide stocks disposed of.

It was the British who were actually the first to use herbicides in a military conflict in the Malayan "emergency."[11,12] Road and rail communication routes are always vulnerable targets in war, and in dense vegetation ambushes are an ever-present threat. To circumvent surprise attacks on their troops the British Military Authorities used 2,4,5-T to increase visibility in the mixed vegetation bordering these routes rather than to cause uniform defoliation.

Earlier trials with the herbicide had shown that this was a reasonable objective. At the same time that the American military authorities were perfecting herbicide delivery systems, British scientists were also reviewing the use of chemicals for controlling plant growth. This was not the first time that British scientists were involved. During the Second World War U.S. scientists are said to have conducted trials in "British tropical stations," and it is known that the scientists in both countries were in close contact.[10] One of the first scientific papers[13] describing the properties of 2,4-D and 2,4,5-T appeared in 1944 (this information was known to the British and U.S. military authorities several years earlier). The paper noted that these synthetic plant hormones (auxins) had potential herbicidal properties. Using this information, two Oxford professors, G. E. Blackman and P. A. Buxton, persuaded the British Colonial Office that the chemicals might have a role to play in controlling the spread of sleeping sickness, a serious medical problem in East Africa. The disease is transmitted by an insect carrier, the Tsetse fly (*Glossina sp.*). The flies thrive in shady habitats, and removal of the shade is thus a standard method of control.

Acting on the scientists' advice, the Colonial Office conducted trials in the early 1950s on the Watusi peninsula of Lake Victoria in Kenya and Tanganyika.[14] The two herbicides, sprayed from light aircraft, which were shown to be effective in inducing leaf fall without killing the vegetation were again 2,4,5-T and 2,4-D. It was not long after this that 2,4,5-T was used on rubber plantations in Malaya to control an outbreak of South American leaf blight, one rubber tree (*Hevea brasiliensis*) in Malaya having been imported from South America. It was only a matter of time before the military application of the herbicide became evident. When that happened, according to Richard Clutterbruck, a former member of staff of the Director of Operations in Kuala Lumpur from 1956 to 1958, it was decided to apply defoliant by hand along road and rail verges and to use aircraft (helicopters, Pioneers, and de Havilland Beavers) for "spraying weedkiller onto enemy cultivation."[12]

The U.S. military's interest in 2,4,5-T revived again with the escalation of the war in Vietnam. In May 1961, the Department of Defense asked the crops division at Fort Detrick to investigate the possibility of defoliating vegetation in Vietnam. Three months later tests were carried out in Vietnam in planes flown by Air Force personnel. From a military standpoint the tests were judged an outstanding success. These final trials were the culmination of a long sequence of tests conducted by U.S. military personnel to effect the eradication of the troublesome water hyacinth[15] from waterways in the United States and the defoliation of trees in Florida, Puerto Rico, Hawaii, and Pran Buri in Thailand. But the Vietnam spraying exercise did not pass without comment. Despite stringent security precautions the test program soon became public knowledge and criticism began to be heard. To allay this reaction, U.S. President John Kennedy sent two senior advisors to South Vietnam to observe the effects of the spraying. The advisers, Walt Rostow and General Maxwell Taylor, were impressed with what they saw and urged that the program continue.[8] Taylor and Rostow's recommendations were accepted and their investigation at first hand eventually brought to an end the U.S. National Security Council's debates on the matter.[16]

The opinion had been expressed in the National Security Council that clearing the vegetation might even be counterproductive, by providing the enemy with a clearer field of fire. But the doubts did not persist. An official spraying program was sanctioned, and Operation Ranch Hand instituted. On 9 January 1962, the first shipment of herbicide was unloaded at Tan Son Nhut Airbase in South Vietnam; the program had begun

Several formulations of herbicide were used in Vietnam, some oil some water based. To avoid cross-contamination and clogging of the fine aerial spray mechanisms, the herbicides were carefully coded.[17] A colored band was painted on each drum for identification purposes. Six colors were used: white, blue, purple, green, pink, and orange. Agents white and blue were the only two water-based preparations; the former, a mixture of 2,4-D and picloram, was used for the long-term control of forest and brushwood vegetation; the latter, cacodylic acid, an arsenic derivative, was a fast-acting preparation used to kill rice and other food crops. The remaining four colors identified oil-based herbicides, all of which contained 2,4,5-T. Agents green, pink, and purple were used until 1964 when they were replaced by the more efficient "Agent Orange" (a 1:1 mixture of the n-butyl esters of 2,4,5-T and 2,4-D). In the early years of the spraying program from 1962 to 1964, 276,000 gallons of herbicide containing 1,692,460 lb of 2,4,5-T was sprayed on the forests of Vietnam.[18] Thereafter the rate of application increased at a considerable pace from January 1965 until February 1971, when the program was stopped. An estimated $11\frac{1}{4}$–$11\frac{3}{4}$ million gallons of Agent Orange were used.[3,18]

For the majority of herbicide missions, C-123 aircraft fitted with 1000-gallon tanks were used. The spray systems were designed to discharge volumes of 1–3 gallons/acre in swathes some 250 feet wide. In the event, all herbicides were applied at the higher 3 gallons/acre (a level ten times greater than that used routinely in the United States). It was reported[18] that the optimum spray pattern could be achieved when "spraying at an airspeed of 130 knots at a 150 foot altitude." Certain operational restrictions were enforced to ensure a degree of uniformity for the spraying. Aircraft were recommended to approach the target just before sunrise with the sun behind them. This procedure afforded some protection from enemy ground fire and took advantage of the existing temperature inversion conditions. Two variables affected a uniform spray pattern—rain, and a wind speed greater than 8 knots. Wind speeds in excess of this occurred on numerous occasions and widened the target area. According to Professor Arthur Westing, Dean of the School of Natural Sciences, Hampshire College, Amherst, Massachusetts, the "essentially unavoidable spray drift" probably accounts for the claim by Vietnamese officials that the area affected by herbicides is greater than that reported by U.S. military sources.[19]

It was in 1964, the year that Agent Orange was introduced to Vietnam, that the first sense of uneasiness concerning the spraying program was voiced in the United States. The Federation of American Scientists issued a statement expressing their concern about the use of antiplant chemicals in Vietnam, claiming that the use of these chemicals suggested that "the United States is using the Vietnamese battlefield as a proving ground for chemical and biological warfare." The Federation scientists registered their opposition to this "experimentation on foreign soil."[20]

Two years later protests about the herbicide program were raised at two other scientific meetings in the United States. A resolution before the June meeting of the Council of the Pacific Division of America's most prestigious science organization, the American Association for the Advancement of Science (AAAS), called for an investigation into the herbicide spraying program. The motion, submitted by E. W. Pfeiffer, associate professor of Zoology at the University of Montana, asked for details of the type of chemicals used, the extent of their use, and their effect, and suggested that a committee be formed to report back on the matter the following year. Council members voted finally 9 to 5 with several abstentions to refer the resolution to the national office of AAAS. The Pacific Division itself they claimed had not the resources to comply with the resolution.[20]

Arthur Galston, Professor of Biology at Yale University, tried equally unsuccessfully to interest his professional association in the herbicide issue. At the 1966 meeting of the American Society of Plant Physiologists, of which he was a past president, Galston attempted to have the subject introduced for debate, but without success. Galston had proposed than an open letter be sent to U.S. President Lyndon Johnson emphasizing that herbicide use on a large scale would cause ecological problems, and that the chemicals may not be as innocuous as had been claimed by military spokesmen. The opportunity to put the proposal never arose; time could not be found for it on the meeting's agenda. Instead, the letter was sent as a personal communication from several individual members of the ASPP, including Galston, rather than on behalf of the Society.[21]

Galston was particularly worried by two aspects of the herbicide issue. Firstly there was the potential for considerable devastation in Vietnam if the herbicides were used on a large scale. Secondly, and of no less importance, was his belief that the U.S. herbicide spraying program was a violation of the 1925 Geneva Protocol prohibiting the use of chemical and biological weapons of war. It was an issue of concern to other scientists too. The strength of feeling was evident when a petition, circulated by Drs. John Edsall and Matthew Meselson of Harvard University, urging the U.S. Government to end its use of antipersonnel and anticrop chemical weapons in Vietnam, was signed by over 5000 scientists.[22] Such was the feeling in the scientific community that 17 Nobel Laureates were among the signatories. The petition was delivered to President Johnson on 14 February 1967. The support for the venture was a personal success for Edsall, then Professor of Biochemistry at Harvard, and the petition represented the culmination of a campaign he had initiated 13 months previously.[23]

In a letter which accompanied the document, some of the signatories set out their views in more detail. They noted that the government had helped to initiate the 1925 Geneva Protocol, but had never signed it. They also pointed out that on 5 December 1966, when the petition was still being circulated, the U.N. General Assembly had adopted a resolution calling on all States to observe the principles and objectives of the Geneva Protocol, and that the United States had supported the resolution. The Government's action was welcomed by the petitioners and it was urged to include herbicides in its definition of chemical and biological weapons. But the Government was unmoved. Johnson did not reply to the petition. The Government continued in its stance that herbicides were not chemical weapons. The spraying program continued. In fact it increased in intensity; in 1967 and 1968, the program was at its peak. It was not until 1975 that the U.S. Government eventually ratified the Geneva Protocol, but the U.S. interpretation of the protocol is that it excludes herbicides.

The unofficial action of scientists to stop the herbicide program in Vietnam entered a more formal phase through the summer and autumn of 1967. The first in a series of articles questioning the morality of the large-scale use of anticrop chemicals[24] and herbicides[25] was published in the journal *Scientist and Citizen*. Meanwhile the AAAS was still grappling with Pfeiffer's much amended resolution from the previous year. Some scientists were equivocal about the resolution, others were opposed to singling out the U.S. action in Vietnam for blame. To accommodate these differing views the AAAS Council finally adopted a more general resolution. It referred to the AAAS' concern about the long-term implications of the use of "biological and chemical agents which modify the environment." The resolution made no reference to the use of these agents in Vietnam. More significantly the AAAS sidestepped Pfeiffer's original suggestion that an investigation be conducted in Vietnam to assess the spraying program, and instituted in its stead an ad hoc committee on environmental alteration.

On 24 and 25 May 1967 the committee, chaired by Dr. Rene Dubos of Washington University, met in Washington's exclusive Cosmos Club. This meeting was a gathering of the faithful; everyone in attendance was a committed environmentalist, but most participants had a different order of priorities for the issues of the day. Few were very strongly moved about herbicide spraying in Vietnam. The report produced by the committee and sent to the AAAS board of directors reflected the different views. Besides recommending replacement of the ad hoc committee by a more permanent commission, it suggested three areas which warranted investigation: those of chemical agents; chemical fertilizers; and waste recycling.[26]

Upon receipt of the report the AAAS directors, at their June 1967 meeting, decided that in view of the complexity of the issues they should be treated in different ways. Thus the more general question of the long-term consequences of environmental alteration was separated from the more specific issue of the use of chemical weapons in Vietnam. The board recognized that no study in Vietnam—an active theatre of war—would be feasible without official government support. A study under official auspices appeared to be the only option. The board therefore instructed AAAS president Don Price to write to the Secretary of Defense to enlist his support for the venture. In his letter to the Secretary, Robert MacNamara, Price outlined the AAAS' concern about the long-term implications of herbicide usage by the military, urged that an immediate investigation of their use be made, and suggested that the National Academy of Sciences–National Research Council "would be an appropriate institution for this purpose."[27]

The Department of Defense (DOD) replied a week later in a letter from Dr. John S. Foster Jr., the Director of Defense Research and Engineering. Foster claimed that the Government had received assurances from qualified scientists that the herbicide spraying program would not have adverse consequences. He continued "Unless we had confidence in these judgments we would not continue to employ these materials." But Foster recognized that there were still uncertainties concerning the long-term effect of defoliant chemicals in Vietnam, and he stated the matter would be investigated.[28]

The Midwest Research Institute (MRI), based in Kansas City, was commissioned by the DOD to produce a comprehensive report on the herbicides detailing what was known about their properties and effects and pointing out gaps in the current knowledge. To ensure that the MRI report carried weight, the DOD asked the National Academy of Sciences (NAS) to comment on the Institute's document. Sections of the MRI report were handed to the Academy for vetting as early as 7 November; by 31 January 1968 the review had been completed. The MRI report was a weighty document, 369 pages in length. Its authors claimed to have contacted 147 individuals for information and listed a bibliography of 1500 references. The

NAS was noncommittal in its review of the MRI's report. The Academy claimed that the MRI had done a "creditable job" in the time available, but noted that most of its citations referred to herbicide use under controlled conditions. What was missing said the Academy was "more factual information on the ecological consequences of herbicide use and particularly of repeated or heavy herbicide applications"; the NAS itself offered to help arrange for the necessary research to gather this material.[29]

Early in February 1968 the AAAS received both the MRI report and the NAS review. Doubts persisted within the Association about how to proceed. Most board members considered the long-term condition of the human environment to be far more important than the use of herbicides in Vietnam. But the board recognized that any committee asked to deal with both issues would be under considerable political pressure to give the Vietnamese situation precedence. By mid-July 1968 the Board of Directors had come to their decision. A policy statement was issued in which the directors rejected the DOD's assertion that there would be no "seriously adverse consequences" arising from the military use of herbicides in Vietnam. The statement, referring to the concern many scientists still shared about the situation, recommended that a field study be undertaken in Vietnam under United Nations auspices to assess the long-term implications for the country.[30] Copies of the statement were sent to the U.N. and to the U.S. Secretaries of State and Defense.

The response from the Secretary General of the U.N. was positive, but restrained. The subject of chemical and bacteriological weapons was receiving "very close attention." Indeed it was. The U.N. would later decide to sponsor an international meeting on environmental quality. As for the issue of chemical and biological warfare, that would be referred by the U.N. General Assembly to the 18th National Disarmament Conference in Geneva for consideration as part of its permanent agenda.[31]

A feeling of resignation was apparent in the DOD's reply to the AAAS statement. John Foster replied again for the Department. He repeated the earlier reassurances he had given about the herbicides and insisted that the DOD was still "confident that the controlled use of herbicides will have no long term ecological impacts inimical to the people and interests of South Vietnam." However, the DOD had an open-minded approach to the herbicide program and would fully support a "systematic scientific investigation" when peace was restored.[32]

The Department of State adopted a more conciliatory approach. Charles Bohlem, a Department Deputy Under Secretary, noted that there were even differences of opinion within the AAAS on the use of certain chemicals. Bohlem recalled that limited studies made by Government agencies had failed to find "serious ecological disturbances" but he acknowledged that only a full investigation "as soon as it was practicable" and after hostilities had ceased, would resolve the matter.[33]

One of the limited studies Bohlem referred to was that of Dr. Fred Tschirley, a Department of Agriculture specialist on tropical ecology. Tschirley's investigation was carried out at the request of the State Department. With only a month from his date of arrival in Vietnam to prepare his report, Tschirley admitted that he could only produce an assessment, and not a detailed investigation of the problem.[34,35] He had had to make his study in the dry season, from mid-March to mid-April, at a time when many trees are naturally defoliated, and at a time which many forestry experts insist is difficult for making an assessment.[36] To complicate matters most of Tschirley's observations were made from the air. Nevertheless his report was clear; the spraying program had caused an ecologic change. The change was not irreversible but recovery, he wrote, may take a long time, perhaps as long as 20 years in the case of the mangrove forests. He stressed that "the desirability of ecologic research after the war cannot be over emphasized." Meanwhile he suggested continuing assessment of the program by aerial survey and ground observation if possible.[34]

Tschirley's report was delivered at a time when the clamor to stop the use of herbicides was becoming an international issue. North Vietnam and China were constantly referring to the use of "poisonous herbicides" in South Vietnam and officials within the Department of State were concerned that this "propaganda" would influence "world public opinion." More seriously, South Vietnamese officials and public were also thought to require some degree of formal reassurance.[37]

This was not long in coming. On 18 September 1968 a press statement was issued by the U.S. Embassy in Saigon on behalf of the Ambassador, Ellsworth Bunker. The statement referred to a "special interagency committee" which the Ambassador had set up earlier in the year to review the herbicide spraying program.[38] The committee had concluded that the program had saved the lives of many "Vietnamese and Allied personnel," that it had undoubtedly created food shortages for the North Vietnamese, and that the economic cost in terms of losses in marketable timber had been significant. It was the committee's opinion that these losses could be mitigated to a large extent by salvage operations and reafforestation and, therefore, it was felt that the benefits of spraying outweighed the drawbacks. It was the opinion of the committee that the defoliation program should continue.

The AAAS Board of Directors remained unmoved by these official investigations. In the Board's view it was clear that an "on-site" investigation, much more detailed than Tschirley's, was of paramount importance. At the annual meeting of the AAAS council, in December 1968, the board announced that the AAAS "would participate in a study of the use of herbicides in Vietnam." The Council, however, would not swallow this. Members objected yet again to Vietnam's being singled out for investigation; the board, they claimed, ought to address itself to other environmental hazards. Anthropologist Margaret Mead insisted that defoliation in Vietnam was "just peanuts" as compared to other environmental hazards caused by technological intrusion such as the building of the Aswan Dam in Egypt. As far as she was concerned, the issue was "warfare, not defoliation."[35]

The issue of Vietnam was dropped from the text of the final resolution, just as it had been two years previously. But the move was not as retrograde as it seemed. The resolution recognized the need for an investigation of the wartime use of herbicides, but also suggested that it would be prudent to consider how the agriculture and economy of the affected areas could be restored. Accordingly the AAAS officials were directed by the Council to prepare specific plans for a field study. The directive had little impact. In June 1969 the AAAS membershp still had no clear indication concerning the ecological questions to be asked and thus had no views on how they might be answered.[39]

Three months previously a number of scientists had tried to find answers themselves. Dr. Egbert Pfeiffer, the original mover of the 1966 resolution at the AAAS on herbicide use in Vietnam, and Dr. Gordon Orians, Professor of Zoology at the University of Washington, obtained $5000 from the *Society for Social Responsibility in Science* to make a brief assessment of the military use of herbicides in Vietnam.[40] Before the SSRS parted with its funds it had to be convinced that the visit to Vietnam would produce worthwhile results; Pfeiffer had no great difficulty in making out a case for the visit although there were those in the SSRS who argued that little of value would be seen on a trip dependent on the help of U.S. military forces.[41]

On March 17 Pfeiffer and Orians flew to Saigon for a 15-day inspection tour of herbicide damage. The publisher McGraw-Hill had agreed to their visiting Vietnam as its own accredited press correspondents[42] on the understanding that their visit would form the basis of a two-article report in the journal *Scientific Research*. U.S. Ambassador Ellsworth Bunker in a telegram to the Department of State in Washington in March 1969 said he was "surprised to learn of the visit" and felt that "nothing new could be learned from it." Bunker asked that the two "be urged to cancel" their trip.[43] But not only did they continue with their plans, they actually received considerable help from embassy military officials in arranging an itinerary.[40] Included in their program was a flight on a spraying mission over the Plain of Reeds, an area northwest of Saigon and close to the Cambodian border. The flight, during which the plane in which they were traveling was repeatedly hit by groundfire, is described in the book "Harvest of Death."[44]

For the two scientists the lessons of the trip were obvious. They admitted that the herbicides were of value to the military and that they did save American lives. However, they had also had a devastating effect on Vietnam's ecology and a detailed investigation was needed to measure the damage.[44] Pfeiffer and Orians admitted privately to being shocked that South Vietnamese scientists knew so little about the herbicides.[40] The major problem had been a lack of information, something which could be remedied, Pfeiffer and Orians claimed, by American scientists supporting basic research by South Vietnamese biologists on the effects of defoliation.[45]

A full account of the visit of the two scientists appeared in the June issue of *Scientific Research*.[46] Up to this point U.S. scientists had been concerned with the threat to the ecology of Vietnam. But in the autumn of 1969 the emphasis shifted dramatically. The reason was the publication of a report claiming that the herbicide 2,4,5-T, used in formulating Agent Orange for spraying in Vietnam, caused malformation in rodents. The study, commissioned in 1965 by the National Cancer Institute for the Department of Health, Education and Welfare was performed by an outside contractor, Bionetics Research Laboratories of Litton Industries. According to the Bionetics study 2,4,5-T was teratogenic in two species of mouse (strains C57BL/6 and AKR) when concentrations of 113 mg/kg body weight were administered to pregnant females on days 6–15 of pregnancy. In the Sprague-Dawley strain of rats, however, only 4.6 mg/kg of 2,4,5-T was needed on days 10–15 of pregnancy to cause a high incidence of kidney anomalies.[47]

The startling evidence that 2,4,5-T was a teratogen, albeit one at a high concentration, was to signal the demise of the herbicide spraying program. Scientists were well aware that defoliants were sprayed over "sparsely" populated areas in Vietnam in concentrations of 3 gallons/acre, ten times higher than that recommended for U.S. 2,4,5-T users. They were also aware that on occasions when enemy groundfire was intense the contents of the 1000-gallon tanks on the C123 aircraft were emptied in 30 seconds,[44,48] as opposed to the normal spray run of $3\frac{1}{2}$–4 minutes.[49] When this information was assembled and presented to Government officials it was not long before restrictions on the use of Agent Orange were introduced.

It was not government scientists who sorted the information, but rather a handful of individual scientists long concerned about the use of herbicides in Vietnam. Matthew Meselson, a Harvard University geneticist, had been given an unofficial copy of part of the Bionetics study by one of his own biochemistry students. The student in turn had received the copy from Anita Johnson, a young lawyer working for the consumer group Nader's Raiders. Johnson had been engaged on a project involving the Department of Health when an official had handed her a copy of the Bionetic's report. The official told Johnson that it was far more important than the project she was then engaged in—an examination of the workings of the Food and Drug

Administration.[50] Knowing Meselson's interest in the issue Johnson passed the report to her student friend to give to Meselson. He in turn had not recognized its significance until a colleague had asked for Meselson's views on reports in the Saigon press claiming that an unusually high incidence of birth defects might be related to the herbicide spraying program. Meselson reread the Bionetics data after having obtained unofficial copies of the full report. The Bionetics study showed 2,4,5-T to be teratogenic. It was possible therefore that the use of the herbicide in Vietnam might be related to the birth defects. Meselson now knew that the government was sitting on a powder keg and that he had the fuse to set it off.

Meselson called for an urgent meeting with President Nixon's science adviser Dr. Lee DuBridge. DuBridge convened the meeting without demur, and agreed to meet a delegation led by Meselson. The delegation included representatives of the Federation of American Scientists and Professor Arthur Galston from Yale, an old friend of DuBridge's. The scientists were well received. Meselson introduced the subject and Galston spoke about the effects of herbicides and about the Bionetic's study in particular. DuBridge had only recently been informed that the Bionetic's study classed 2,4,5-T as a teratogen. His meeting with Meselson's team convinced him that he would have to act and he did so by announcing publicly that restrictions on the use of 2,4,5-T would be introduced in the U.S., and in Vietnam the Department of Defense would "restrict the use of 2,4,5-T to areas remote from the population."[51] As Thomas Whiteside points out in his book on 2,4,5-T *The Withering Rain*, DuBridge's statement was much hedged about with qualifications.[52] Referring to the large doses of 2,4,5-T required to produce malformations in the rodents studied and the small number of animals used in the investigation, the statement said that it was "improbable" that individuals could receive harmful effects from "existing" uses of 2,-4,5-T. The statement added that further evidence on the herbicide was being sought.

On the face of it, DuBridge's statement implied a restriction on the use of 2,4,5-T in Vietnam. But this was not the case. On the following day October 30, the Department of Defense announced that there would be no changes in the military uses of 2,4,5-T—the armed services were already conforming to the DuBridge directive.[53] As far as the Department of Defense was concerned herbicides were a valuable part of its armory. Rear Admiral William Lemos of the Policy Plans and National Security Council Affairs Office of the Department of Defense summed up the Department's view in testimony before a subcommittee of the House of Representatives Foreign Affairs Committee. "We know . . . that the enemy will move from areas that have been sprayed. Therefore, enemy base camps as unit headquarters are sprayed in order to make him move to avoid exposing himself to aerial observation."[53]

This apparent intransigence on the part of the DOD, coupled with the

DuBridge statement and the results of the Bionetic's study, was enough to finally sway the AAAS membership at its annual meeting in December 1969 to vote for a resolution to carry out an investigation of herbicide damage in Vietnam as originally proposed 2½ years previously. Meselson was asked to conduct the investigation and to prepare a plan for the one-year program. The AAAS agreed to underwrite the study and to provide $80,000 for what became known as the Herbicide Assessment Commission.[54] Arthur Westing, then Professor of Botany at Windham College in Vermont, conducting an investigation of herbicide damage in Cambodia at the time (see Chapter 8) was appointed by Meselson to direct the work of the HAC.[55] The Commission began its work in February 1970 and by May had produced a plan of campaign. Thereafter events moved swiftly. In June a five-day working conference was held to discuss possible problems and areas of investigation for the HAC visit. Twenty-three specialists with expertise in a range of subjects attended the briefing. A month and a half later at the end of July, everything had been arranged and the Commission left for Saigon.

In the months that the HAC had been deliberating about its proposed investigation government agencies had been reviewing the Bionetic's Laboratory findings. By April 1970 the reviews had been completed. Some action on 2,4,5-T was considered to be essential. In a joint statement the Secretaries of Agriculture, Health, Education and Welfare and the Interior announced the suspension of certain uses of 2,4,5-T.[57] The Department of Defense subsequently followed suit and announced that it too had suspended the use of the 2,4,5-T-based Agent Orange.[57] But the DOD's orders seem to have taken rather a long time to reach all its units. According to a New York Times report, Agent Orange had been used in two northern provinces of South Vietnam throughout the summer of 1970.[58]

The removal of Agent Orange from the official herbicide armory was an auspicious start for the HAC. The four-man Commission—it now included John Constable, Professor of Surgery at Harvard Medical School, and Robert Cook, a graduate student in ecology at Yale University—arrived in Saigon on 1 August to start their three-week investigation. Although the members of the Commission were in Vietnam officially as private citizens, the Department of State had asked government agencies in Vietnam to "bend a few regulations" to accommodate the visitors. Ambassador Bunker met the scientists three days after their arrival and offered his cooperation. To facilitate their mission Bunker enlisted the help of General Rossyn, acting commander of the Military Assistance Command for Vietnam. According to Meselson, Rossyn helped considerably, and provided the team with a helicopter.[59] Meselson knew that the cooperation of the military was vital if their investigation was to be successful. To avoid antagonizing officials in Vietnam, Meselson made it clear that the HAC should have no contact with the press or with any Americans with pronounced "antiwar" views. Contact of this sort Meselson feared might be used to discredit the team's findings at a later date.

The Commission left Saigon for the United States on 23 August again with instructions from Meselson to have no contact with the press; there would be plenty of time for comments when the final report would be presented at the AAAS annual meeting in 1970. When the Commission had finished its analysis it too would report that there had been considerable devastation to hardwood and mangrove forests as a result of herbicide spraying. The crop destruction program—authoritatively estimated to have destroyed food capable of feeding 600,000 people for a year—largely confined to the central Highlands of South Vietnam was thought by the Commission to have had a profound impact on the one million Montagnard hill people resident in the area.

But the Commission's most controversial findings were those on stillbirth rates and the incidence of malformation. An earlier study carried out by Dr. Robert Cutting on behalf of the Department of the Army and the (South) Vietnamese Ministry of Health had concluded that there was a decrease in the national rate of stillbirths for the period 1960–1969. Cutting's team divided the ten-year period into two parts to represent the years of light and heavy spraying of herbicides. For the period 1960–1965 (light spraying) the stillbirth rate was 36.1/1000 live births, for the years 1960–1969 (heavy) it was 32.0/1000.[60] There was thus a decrease in the rate of stillbirths at the same time that herbicide spraying had increased. The Herbicide Assessment Commission viewed the data in a different light. The Commission argued that some two-thirds of the births studied by the army had occurred in the Saigon area, the region least affected by herbicide spraying. If then, the figures for the capital area were subtracted from the total, the downward trend of stillbirths reported by Cutting was reversed. The stillbirth rates for the provinces became 32.0 for 1960–1965 and 38.5 for 1966–1969.

If the same revision was applied to the figures for congenital malformations, Cutting's downward trend for the two phases 1960–1965 and 1966–1969 was again reversed. The Commission noted, however, that the revised malformation rates 2.3/1000 live births for 1960–1965 and 3.1/1000 for 1966–1969—much lower than those for Western Europe—were still exceedingly low. In their own investigation the Commission had been told by hospital midwives that deformities were often seen but never recorded[55]; many Vietnamese regarded deformed children as a stigma.

Meselson's team reported two other significant findings. The average rate of stillbirths for Tay Ninh City Provincial Hospital—representing an area heavily sprayed with 2,4,5-T—for 1968/1969 was 68/1000 live births, double the 31.2/1000 figure in the Army sample for the whole country.[55]

As for the incidence of specific types of deformities the Commission reported that there had been a major increase in the incidence of spina bifida and cleft palate in 1967 and 1968 at the Saigon Children's Hospital. The hospital dealt with referral cases just over half of which were from the area of the capital, Saigon. Neither the Commission, nor the doctors at the Hospital concerned, could explain the increase in malformation, and both argued that more detailed research was required to identify the provincial origin of cases and possible herbicide exposure.[55]

Before presenting his team's findings to the AAAS 1970 convention in Chicago, Meselson informed officials at the Department of State and the White House of the outcome of the HAC's investigation. Department of Defense officials ignored Meselson's offer of an advance briefing. However, the military commander in Vietnam, General Creighton Abrams, and Ambassador Ellsworth Bunker had already been informed of the Commission's findings on defoliation and crop destruction.[59] Abrams and Bunker were thus prepared for the White House statement on 26 December 1970 that the military use of all herbicides in Vietnam was to be rapidly phased out.

This White House statement was an attempt to preempt the publicity expected for the HAC's report to the AAAS convention. It had only marginal impact.[61] Meselson and Westing presented their team study to a glare of publicity from interested news media. In the convention itself delegates were both moved and enraged by the HAC's findings. AAAS council members were similarly affected, and argued that the herbicide spraying program in Indo-China should be halted immediately. A resolution to that effect, put forward by E. W. Pfeiffer, was passed by the Council with overwhelming support. As a mark of appreciation the Council adopted a resolution thanking Meselson and his colleagues for carrying out the investigation and also thanking Pfeiffer for his persistent efforts in urging that the AAAS undertake the study.[62]

On 20 February 1971 Ambassador Bunker and General Abrams, in a statement issued from Saigon, announced the virtual termination of the herbicide program.[5] In future, spraying operations would only be used to clear vegetation around "allied" (U.S. and South Vietnamese) base camps and in the manner authorized for the U.S. use of 2,4,5-T by the Department of Agriculture.[5] In September 1971 the Department of Defense ordered the return of existing stocks of Agent Orange to the U.S. for disposal in an "environmentally safe" and efficient manner.[63] When the use of Agent Orange had been suspended in 1970 the U.S. Air Force had an inventory of 1.37 million gallons in South Vietnam and 850,000 gallons stored in Gulfport, Mississippi. In April 1972, the South Vietnam stock was moved to Johnston Island in the Pacific for storage.

At first, the Air Force was reluctant to dispose of the 2.2 million gallons of herbicide. The original purchase price had been $16.5 million. By

1974 it was valued at about $80 million and represented a valuable asset. The Air Force considered selling this surplus stock to South American governments for their domestic use. Many of the governments were only too eager to purchase the product; however, adverse publicity in the United States forced the Air Force to abandon the idea. It tried instead to register the herbicide preparation for sale on the open market in 1974.[65]

This plan also proved to be unsuccessful. Opposition was so widespread that the Air Force finally realized that disposal of the herbicide was the only course open to it. Various options were considered, including incineration on Johnston Island or in the United States itself. The only feasible proposition however, according to the U.S. Environmental Protection Agency, was incineration at sea. The specially constructed Dutch vessel "Vulcanus" owned by Ocean Combustion Services of Rotterdam[66] was commissioned for the project. In the summer of 1977 all remaining stocks of Agent Orange were burned. Vulcanus did not escape unscathed from the program. Small amounts of dioxin contaminated the vessel during the incineration, necessitating an extensive cleaning program.[67]

Criticism of the herbicide program had had an impact on the Department of Defense before Meselson's report back to the AAAS in December 1970. As far as herbicide usage in Vietnam was concerned, it was clear to many that the DOD was the villain of the piece. Officials in the Department had always insisted that the herbicides used in Vietnam were safe. The Bionetics study showing 2,4,5-T to be teratogenic, coupled with the Pfeiffer and Orians study of herbicide damage and the AAAS investigation, had shattered this complacency. U.S. Congressmen and Senators were similarly shaken. It was clear that a full investigation was required to answer outstanding questions. Legislation requiring the DOD to finance a large-scale scientific study of herbicide damage in Vietnam—inserted as a provision of the 1970 military authorization bill—was enacted in October 1970. A contract authorizing the National Academy of Sciences to undertake the study was signed on 8 December 1970 by DOD and NAS officials. Originally planned as a one-year project,[68] the NAS study would take more than three years to complete; it was published eventually in March 1974.

Before 1970 the fears expressed about the effects of herbicides had centered on their military use in Vietnam. In this context, the chemical causing most concern at first was the crop-destroying Agent Blue, an arsenic-based compound. In 1969 the emphasis changed, and 2,4,5-T became a cause for concern. The change also propelled 2,4,5-T into the international arena; its use became a controversial issue in many countries. The controversy was fueled by the suggestion in 1970 that the teratogenic properties of 2,4,5-T might be due to its contamination with 2,3,7,8-tetrachlorodibenzo-p-dioxin (dioxin). Dow Chemical, the principal U.S. producer of 2,4,5-T, had known about the presence of a highly toxic impurity in the herbicide since 1950.[69]

By 1965 Dow had identified the impurity as dioxin and informed other herbicide manufacturers of its finding. An accident the previous year in its 2,4,5-trichlorophenol plant—the starting point for 2,4,5-T production—had caused an outbreak of chloracne the following year. When the company rebuilt its trichlorophenol plant in 1966, dioxin contamination of 2,4,5-T produced was kept below 1 ppm.

Agent Orange, the 2,4,5-T formulation used in Vietnam, was not so clean. Residual stocks of Agent Orange on Johnston Island, when analyzed for their dioxin content, were found to have concentrations of the contaminant ranging from 0.05 to 47 ppm, with an average value of 1.91 ppm. At least $11\frac{1}{4}$ million gallons of Agent Orange were used in Vietnam and it has been estimated, therefore, that some 220–360 lb of dioxin was dispersed over the country.[70] Arthur Westing, coleader of the AAAS Herbicide Assessment Commission, has claimed that this estimate may be too low.[4] The residual herbicide analyzed for its dioxin content was manufactured in the late 1960s when processes to control dioxin contamination had been improved. Westing believes that, had samples from the mid-sixties been available for analysis, measurements may well have revealed higher dioxin values.

Preparations of 2,4,5-T with a dioxin content of 30 ppm were shown to cause malformations in rat fetuses in studies published in 1970.[71] In the following year dioxin itself was confirmed as a teratogen,[72,73] and 2,4,5-T formulations with low dioxin contents (0.5 ppm) were found not to cause malformations in the same species of animals affected by the more contaminated 2,4,5-T the year previously.[72]

It was thus now firmly established that dioxin contamination of 2,4,5-T was a major problem. Steps were clearly required to reduce the level of contamination and in March 1971 a report from the President's Science Advisory Committee recommended rigorous control to limit dioxin in 2,4,5-T to not more than 0.5 ppm.[74] The Advisory Committee advocated lowering it even further to a limit of less than 0.1 ppm. The 0.1-ppm recommendation was supported two months later in the report of the Advisory Committee on 2,4,5-T to the Administrator of the Environmental Protection Agency.[75] Formed in 1970 to control pollution of the environment, the EPA inherited the 2,4,5-T problem from the Department of Agriculture. Ten years later it would still prove to be a very live issue for the Agency to deal with. By 1979 the EPA had become convinced that even the 0.1-ppm dioxin level presented a health hazard and in 1980 the Agency began hearings which it believed would support its supposition.

The controversy over the domestic use of 2,4,5-T would eventually spread to many countries but the arguments were never so heated, nor the stakes so high, as those in the United States. Loss of 2,4,5-T sales revenue was, and remains, a problem for U.S. chemical companies. However, these sums pale into insignificance when compared with the compensation claims against the company by individuals who insist that they have been injured by the herbicide.

The decade 1970/1980 was witness to repeated attempts by the EPA to have 2,4,5-T withdrawn from the market. But every attempt was foiled by the Dow Chemical Company. Dow was able to argue repeatedly in court or at EPA hearings that there was insufficient evidence to support the allegations that dioxin in 2,4,5-T presented a health risk.[76] Each defeat left the Agency undaunted. The 1980 hearings are testimony to its persistence; the Agency claimed in its prehearing submission that there are now sufficient data confirming that 2,4,5-T is indeed dangerous.[77]

Concern about the effects of the military use of herbicides became an international issue in 1972. The U.S. Department of State several years previously had taken steps to minimize the repercussions of the Vietnam spraying program in the international arena (see p. 156). However, these efforts did little to allay the fears of Sweden's Prime Minister Olaf Palme. In his opening address to the 1972 U.N. Conference on the Human Environment in Stockholm, Palme denounced the large-scale use of herbicides in warfare. Referring to the destruction brought about by this use as "ecocide," Palme said that it was of paramount importance that "ecological warfare cease immediately."[78] Without referring directly to the U.S. spraying program in Vietnam, the Swedish Prime Minister's tough words clearly implied criticism of American activity in Indo-China. At the last minute Palme departed from his prepared speech to insert a section denouncing the use of chemical and biological weapons. On the evening before his speech Palme had asked to see Pfeiffer and Westing, both of whom were attending the conference, for a briefing on the effects of herbicides.[41,79]

The reaction to Palme's speech was, perhaps, predictable. A denunciation followed immediately from Russell Train, chairman of the U.S. delegation to the U.N. conference. Train made it clear that he was speaking on behalf of the U.S. government, which objected strongly to what it considered to be the gratuitous politicizing by Palme of the environmental discussions.[80] The Swedish Prime Minister continued to criticize the U.S. involvement in Vietnam on different occasions throughout 1972. His comments finally proved too much for the Nixon Administration and the U.S. Ambassador to Sweden was recalled to Washington. This low point in diplomatic relations between the two countries was to last from January 1973 until May 1974 when Ambassadors were once again exchanged.

By the time the National Academy of Sciences published its report in March 1974 on *The Effects of Herbicides in South Vietnam*,[81] it was well known that Vietnam had suffered widespread damage as a result of the herbicide spraying program. The NAS study confirmed much of what had already been reported by earlier investigators. An estimated 36.1% of the mangrove forests had been sprayed and large contiguous areas were devastated (see Figure 14). Unless an urgent replanting program was instituted, the NAS report said, the region could be barren for over a century, perhaps longer. The Academy estimated that 10.3% of inland forests had been sprayed one or more times, causing the loss of 1.25 million cubic meters of marketable timber.

Not everyone on the NAS committee agreed with these estimates of timber loss. Two members of the team—Professor Pham-Hoang-Ho from Saigon University and Professor Paul Richards of the University College of North Wales (U.K.)—dissented from the report, on the grounds that it underestimated these losses. Ho claimed that the report's estimate was ten times too low. Professor Meselson, who was one of the members of the report's review panel, also said that the estimate of tree damage was wrong by a factor of ten. Meselson pointed out that the report's figure for the density of trees in Vietnam's forests was considerably less than the density of trees on the Academy's own front lawn, a lush, very lightly wooded expanse of grass. Meselson argued that there was clearly something wrong with the interpretation of the aerial photographs used to assess the damage and he called for a reassessment.[82] His call was ignored. When Meselson later at-

FIG. 14. Dead, herbicide-sprayed mangrove trees, South Vietnam. Photo A. Westing.

tempted to obtain the photographs to make his own assessment he was informed that the 35,000 photographs had been returned to the Department of Defense and destroyed.[59] Dr. Thomas Dashiell, staff specialist for chemical technology at the DOD told the author that "the photographs had caused physical storage problems. In consequence they had been destroyed in late 1977 through the spring of 1978." It seems, however, that although some probably irreplaceable scientific evidence has been destroyed, all is not lost. According to Dashiell, when the photographs were destroyed, the silver was recovered in accordance with "legislative mandates"![83]

As to the effect of herbicides on the health of the Vietnamese population, the NAS report was quite specific. Interviews with the Montagnard people, conducted by Dr. Gerald Hickey of Cornell University, provided remarkably consistent reports of the range of effects on children following spray missions; diarrhea, skin rashes resembling insect bites, abdominal pain, and death.[82]

The NAS report also stated that it could find no conclusive evidence of association between exposure to herbicides and birth defects in humans.[84] Unable to perform its own investigations in Vietnam because of the fighting, the NAS team subjected the 1970 study of the Herbicide Assessment Commission to a rigorous examination. The increased incidence of spina bifida and cleft palate reported by the HAC in the years of heaviest herbicide spraying was noted. However, the NAS group thought the records were insufficient and hence no firm conclusions could be drawn.

In 1973 Meselson and a colleague, Robert Baughman, developed a highly sensitive assay for dioxin and reported that they had detected the chemical in fish and shellfish taken from rivers draining herbicide-sprayed regions in South Vietnam.[85] This evidence confirming that dioxin had entered the food chain in Vietnam worried some of the NAS team members. They considered that it implied a health risk for the Vietnamese population. This concern was reflected in their report, which mentions the information as a "matter that warrants further attention."[86]

Some of the report's more scathing comments were reserved for a 1971 report from North Vietnamese scientists that refugees from South Vietnam exposed to herbicides had a higher incidence of damaged chromosomes.[87] The Vietnamese scientists claimed that refugees directly exposed to herbicides or residents in areas which had been sprayed had a higher incidence of chromosome abnormalities—breaks and gaps—than a control population. The NAS report said that this study was statistically inadequate, and that it had not considered other agents such as viruses which might cause chromosome damage. Furthermore, breaks and gaps, the NAS report claims, were not the abnormalities expected from exposure to chemicals months or years previously. And finally the NAS reviewers queried the figure of 0.4% for the incidence of chromosome breaks and gaps reported in the Vietnamese control population. The comparable rate for control populations in the West and Japan is of the order of 1.0% and often higher; the 1% figure was the one the Vietnamese scientists reported for their herbicide victims, and one they considered to be abnormal.

These criticisms have not gone unchallenged. The Vietnamese scientists involved claim that their study was statistically inadequate because it was performed during the War when they had no opportunity to do a controlled study. They insist that their work has been checked for accuracy by other Western scientists. Dr. Bach Quoc Tuyen, a cytologist and one of the authors of the 1971 paper, was challenged on the question of discrepancies in the values of breaks and gaps reported for control populations in Vietnam and the West. To what did he attribute the higher 1.0% figure for the West? "Chemicals. Perhaps a higher background level of radiation."[88] Western scientists remain skeptical about Tuyen's claim. Further studies are clearly needed to resolve the existing differences of opinion on this issue.

Chromosome damage is not the only medical problem Vietnamese scientists attribute to exposure to the 2,4,5-T based herbicide Agent Orange. Professor Ton That Tung, a surgeon and director of the Viet Duc Hospital in Hanoi, blames the dioxin which was present in Agent Orange for the current increase in primary liver cancer in Vietnam.[89] Tung admits however, that the evidence is only circumstantial and that there is no good epidemiological data to support his thesis.[88] There is a real increase in cancer of the liver in South Vietnam, Tung claims, an observation which he says is not true for North Vietnam. Agent Orange was only used in the south of the country; Tung suspects that the herbicide's dioxin contaminant could be responsible for the increase in cancer cases. As he is the first to point out, this is still supposition.[90] Western scientists remain skeptical. They point out that most known carcinogens have a latency period of 20–30 years between exposure to the the carcinogen and the development of tumors; there was a shorter period of time between the deposition of dioxin in Vietnam—beginning in 1962—and the appearance of liver tumors.[88]

In a recent unpublished study[91] Tung and two of his colleagues claim that former Vietnamese veterans of the War exposed to herbicides have had children with an abnormally high incidence of congenital malformations. Tung et al. claim that the effects of herbicides are transmitted from the father to the first generation of children. Soldiers serving in South Vietnam and exposed to herbicides for 3–4 years were compared with colleagues who had only served in North Vietnam. Both groups were married to women who had never left the North. The 650 soldiers who had been south fathered 1158 children in the four-year period 1975–1979. Of the 1158, 43 were born malformed. The 114 soldiers who remained in the North fathered 309 children; no abnormalities are reported.

Among the 43 malformed children two types of malformation were more common than others. Fifteen of the 43 had congenital heart disease and nine had defects of the nervous tissue involving the brain and spinal cord. The overall incidence of malformations observed in the study group is 3.6%, higher than the expected average of 1.5%–2% the authors expected to find. However, the expected average for Western Europe and the United States is higher than this. Major malformations (spina bifida, cleft palate, Down's syndrome, etc.) account for 2%–3% of live births in these countries. This figure rises to 3%–7% if major malformations of the cardiovascular system (including the heart), respiratory tract, gastrointestinal tract, genitals, and urinary tract are included.[92]

This latest study from Vietnam has not yet been thoroughly evaluated. Initial reactions to it from some Western scientists are cautious. They say that the more usual actions of toxic chemicals is to kill the sperm outright. And when sperm is damaged but not killed, it is considered more likely that the fetus will abort. It is possible, but unlikely, that dioxin could damage sperm leading to a major developmental defect in the fetus.[93]

Animal experiments lend little support to this last possibility, however. A recently completed study by Dr. James Lamb for the U.S. National Toxicology Program reported that simulated mixtures of Agent Orange containing high and low concentrations of dioxin fed to male mice had no effect on their fertility or reproductive ability. The dioxin-treated males were mated with untreated female mice. The survival and development of offspring born to these animals was unaffected by paternal exposure to the chemicals. Malformation rates were variable in the four animal groups (one control and three dioxin-dosed groups) used in this study. When observed over an eight-week period, there was no uniform pattern discernible in the variations, consistent with exposure to the chemicals. As for the incidence and types of malformation observed, these were similar in all four groups.[94]

Lamb's study was initiated in response to claims by former U.S. servicemen that exposure to Agent Orange had damaged their health. Exposure to the herbicide, the Vietnam veterans claim, is the cause of their children being born with birth defects, and responsible for their own loss of libido and fertility.[95–99] By 15 September 1979 417 veterans had lodged complaints. Half of the men reported skin problems and a quarter said they experienced nervousness, headaches, and fatigue. Other illnesses recorded by Lieutenant Colonel William Wolfe of the Air Forces Epidemiology Division included peripheral nerve damage, gastrointestinal and genitourinary complaints, cancer, respiratory problems, ear, nose, and throat troubles, and cardiovascular problems.[98] Some 190 veterans claim that they had been exposed to the herbicide, but had experienced no illness as a result.[100]

The veterans' complaints were first aired publicly by a CBS-affiliated television station in Chicago on 23 March 1978. The hour-long program suggested that the veterans and their children had a variety of health complaints which might be associated with herbicide exposure. In response to these allegations, the U.S. Air Force Surgeon-General in April 1978 asked the Air Force's Occupational and Environmental Health Laboratory to assess the complaints. The laboratory's 247-page report was issued in October 1978. Acknowledging that long-term effects of exposure to dioxin were not yet known it concluded: "In the absence of chloracne (a positive sign of exposure to dioxin), systemic symptoms would have been unlikely to occur. It is conceivable that mild chloracne symptoms could have developed undetected and mild systemic symptoms such as central nervous system dysfunction, peripheral neuropathy (nerve damage) or other systemic involvement could have also gone undetected. These symptoms, however, would have cleared shortly after removal from exposure and any current symptoms are most probably due to some etiology other than the past exposure of the individuals to Herbicide Orange in Vietnam."[101]

This negative verdict left the veterans undeterred. As more veterans came forward their cases were taken up by Victor Yannacone Jr., an attorney with previous experience of chemical contamination cases. Yannacone was initially asked to take up the case of Paul Reutershan, a former helicopter crew chief in the Vietnam War. Reutershan had flown on several herbicide spraying missions during which Agent Orange was used and he believed that the cancer of the colon which he had developed was a result of this exposure. The crew chief died in December 1978, after Yannacone had filed a suit for damages against six chemical companies, including Dow Chemical. Following his death, a new suit was brought on Reutershan's behalf and on that of "all American servicemen whose health has been damaged because of contact with Agent Orange."[76]

In May 1979 the Carter Administration ordered that a wide-reaching investigation of the servicemen's complaints be undertaken by the Veteran's Administration (VA).[102] The initial reluctance of the Air Force to countenance such an investigation gradually changed until in 1980 Lieutenant Colonel Wolfe announced that clinical evaluation alone would be insufficient to determine a cause and effect relationship "between abnormal health status and herbicide/dioxin exposure." A detailed epidemiologic and statistical study would be required before this evaluation could be made, said Wolfe. The purpose of the diagnostic program planned by the VA, and likely to take several years to complete, was not to "condemn or defend the use of herbicides in the Vietnam war, but rather it is to identify any adverse health effects in personnel who were exposed to those chemicals and contaminants."[103]

Yannacone's main concern is for those veterans who have developed cancer and for those who have fathered children with congenital birth defects. He admits to having some difficulty in confirming that veterans have dioxin poisoning; the skin disease chloracne is a positive indication of this. However, Yannacone says that it is difficult to find expert dermatologists capable of distinguishing chloracne from "jungle rot"[104] (diphtheria of the skin caused by bacteria) and other skin diseases. VA officials on the other hand, claim that this identification is not at all difficult and offered to study Yannacone's patients to confirm the diagnosis of chloracne. Yannacone, however, was not prepared to hand over too much information to the VA. Many of the lawyers representing the veterans in their claims against the companies distrusted the VA, and claim that it was at first unwilling to concede that U.S. servicemen might have a case against the herbicide. The VA, they say, is only now investigating the situation because it has been forced to do so. For the moment, Yannacone's first task is to establish that the chemical companies knew about the presence of dioxin in the 2,4,5-T they sold to the U.S. government for use in Vietnam, and then to ascertain whether or not they knew that dioxin was toxic. The evidence would suggest that Yannacone will have little difficulty in establishing this as Dow was aware of the presence of a toxic impurity in the herbicide as far back as 1950. By 1965 this had been identified as dioxin (see p. 163).

Of the $2\frac{1}{2}$ million U.S. servicemen who served time in Vietnam some 30,000 had presented themselves to VA hospitals by October 1980 for inclusion in the Administration's program. The results of their medical investigations are to be analyzed by computer to investigate any potential relationship between their state of health and the use of Agent Orange. Officials in the VA admit privately that it will prove difficult to determine this relationship as the extent of the men's exposure to the herbicide is not known. The VA is pinning most of its hopes of finding any problems arising from the use of herbicides in Vietnam on a study of the 1200 Air Force personnel who dispensed the chemicals over Vietnam on behalf of Operation Ranch Hand. Exposure of this much smaller group is easier to document.

Birth defects in children of U.S. servicemen known to have been in Vietnam are also to be investigated for any causal connection with Agent Orange. Some 12,000 children with birth defects, 6000 of them major malformations, are being studied at the Center for Disease Control in Atlanta, Georgia. The children's parents are to be traced and details about their past history, particularly their time spent in active service in Vietnam, noted.

In addition, the U.S. Air Force Institute of Pathology at the Walter Reed Army Hospital in Washington is to compile a cancer registry for former veterans of the Vietnam War. The incidence of tumors, particularly soft tissue sarcomas, will be recorded in an attempt, again, to observe whether a causal relationship exists between the incidence of cancer and exposure to Agent Orange.

These epidemiological studies could assist Yannacone's case. Meanwhile he is determined to seek redress for the clients he represents. By July 1980, he had more than 2000 compensation claims against the chemical companies ready to present to court. A further 3000 claims were being reviewed by law firms cooperating in the prosecution of the companies.[104]

Ten months later in May 1981, the number of petitioners had increased substantially. The number of veterans who had filed as plaintiffs through Yannacone had risen to 3500 and the number represented by other law firms stood at 11,500.[105]

The Air Force acknowledges that individuals attached to the Ranch Hand squadron and to individual helicopters used for spraying were the servicemen most likely to have come into contact with herbicides.[106] However, in a letter to Senator Charles Percy, the Department of Defense admitted that "the chances that ground troops were exposed to Herbicide Orange are higher than the DOD previously acknowledged."[107] Yannacone's potential claimants also include some 1500 Australian veterans of the Vietnam War.[108] In view of their complaints, the Australian Government is to conduct its own two-year investigation to determine whether the incidence of birth defects in the veterans' children is higher than average.[109]

On 7 August 1980 Yannacone filed the first $150 million suit against the chemical companies on behalf of 75 veterans.[110] The final suit could involve a total of $40 billion.[107] But the chemical companies may yet escape with little but a tarnished image. The companies, which include Dow Chemical, Monsanto, Thomson Hayward, Hercules Incorporated, and Diamond Shamrock, are looking to the U.S. government for reimbursement of any damages they may have to pay. The companies insist that they were only acting on behalf of the government and supplied what was asked of them. They also claim that they had no control over the use to which herbicides were put in Vietnam; this they say was entirely the responsibility of the government, and hence if the veterans are to receive compensation they should be paid out of public funds.[108] The case being brought by the veterans is one of the most expensive in American legal history. It could also prove to be one of the longest cases on record. Part of the problem is due to the complexities in the law between different States. According to an article in the *National Law Journal*, "if New York law applies to the litigation, for instance, a tort action must be filed within two years of the occurrence of the injury. Under Illinois law, for example, a tort action may be filed within two years of the 'discovery' of the injury."[111] In some states the action must be brought within two years of finding the actual *cause* of the injury.

To overcome this complexity, Yannacone and his associates attempted to have the case heard under Federal Common Law. This would have meant that one law would be used in all states, and that all veterans would be included in the action irrespective of the time of onset of their illness or the date at which they suspected herbicides were to blame. But the lawyers failed in their attempt. The second U.S. circuit Court of Appeals in New York ruled in December 1980 that the case could not be tried under Federal Law.[112]

Three weeks later, the veterans won a small victory when a Federal court in New York ruled that they need not apply individually for compensation but that they were entitled to pursue a class (joint) action against the chemical companies. Giving his ruling, Judge George Pratt said that it was only the companies who could be sued and not the U.S. Government. The injuries the veterans received, the judge ruled, were "incidental to military service."[113] Following these rulings, the veterans' lawyers were preparing their case for the courts.

The suit will be heard in phases. The courts must first decide whether the Government or the companies are liable for damages. They will then consider whether the companies knew that Agent Orange was heavily contaminated with dioxin when it was used in Vietnam. Finally the courts will rule on whether or not dioxin was the cause of the veterans' medical problems.[105]

The possible consequences of exposure to dioxin-contaminated herbicides are also being considered in another court where the Environmental Protection Agency is arguing the case for the banning of 2,4,5-T. If the current EPA hearings on 2,4,5-T provide sufficient evidence to show that the herbicide poses a health risk and should be banned, Dow Chemical stands to lose $12 million in sales and will undoubtedly face a string of lawsuits by claimants alleging that 2,4,5-T has affected their health and that of their children.

The EPA hearings opened on 15 March 1980 and were expected to run for 18 months to 2 years, before the supervising judge, Edward B. Find, gave his verdict. The EPA administrator would then issue the final ruling on whether or not a ban should operate on the use of all 2,4,5-T-based products. 150 witnesses were to be called by the EPA to support its case that 2,4,5-T should be banned. Dow Chemical had promised to invalidate most of the EPA's case and to prove in the words of its Vice-President and health director Etcyl Blair that the herbicide was "truly a safe product, a very useful product for which there is a real need."[114]

On 28 February 1979 EPA's administrator Douglas Costle issued an emergency order suspending the use of 2,4,5-T in forestry, rights-of-way, and on pasture. The grounds for suspension was an EPA-commissioned study which had found that the use of 2,4,5-T was linked to an increase in spontaneous abortions in the State of Oregon. Costle issued his order to prevent the new season spraying then about to begin.[115]

The EPA commissioned Dr. Eldon Savage of the Environmental Studies Program of Colorado State University to investigate allegations by women residents in Oregon that a spate of miscarriages could have been caused by exposure to 2,4,5-T. Savage's investigation, "Alsea II,"[116] found a strong correlation between the use of the herbicide and a tripling in the rate of spontaneous abortions among local women. A 1600-square-mile rural area where 2,4,5-T spraying had occurred was chosen as Savage's study area. A 1000-square-mile rural area and an urban area were chosen as controls. Evidence on spontaneous abortions in the first 20 weeks of pregnancy was taken from the records in those areas.

According to the EPA Savage's study revealed that the spontaneous abortion index (ratio of abortions to live births) was higher in the study area than in the rural area used as a control. For the months of May, June, July, and August, the index for the study areas was 89.9, 130.4, 105.4 and 88.1 compared with 63.2, 46.0, 55.3, and 79.8 for the rural control. In the urban area the ratios were even lower. The investigation showed a significant correlation with the 2,4,5-T spray pattern, with a lag period of two to three months. The residents, the study suggests, could have come into contact with the herbicide and its dioxin contaminant in drinking water or through eating fish and other wildlife.

Critics of the EPA study—and there are many—dispute the existence of evidence to show that there is a relationship between 2,4,5-T spraying and spontaneous abortions. One critic, Professor Nathan Mantel, a statistician at the biostatistics center of George Washington University, Bethesda, claims that there are "fundamental statistical and logical flaws" in the report of the EPA study.[117] Mantel was asked to review the report by the National Forest Products Association. He says that the EPA should not have included an urban control area in its program and compared results from this with those obtained from the rural study area. The EPA later downgraded the importance of the urban control area for its case against the herbicide. Mantel's analysis revealed no statistically significant differences between the data collected from the EPA rural "study" area, and a third "control" area which Mantel selected, and one which he considered offered a more suitable comparison.

Referring to the Alsea findings of a correlation between spray times and a seasonal pattern of abortions, Mantel insists that the comparison was based on an improper statistical approach. He claimed this result was of little value anyway, since there were no statistically significant differences in the overall rates of abortion in the areas investigated; the EPA study reported little difference in the percentage of hospitalized spontaneous abortions for women aged 20–49 in the three areas.

These are not the only criticisms leveled at the EPA's report. It has also been pointed out that the June peak in miscarriages occurred in only one of the six years under investigation; spraying occurred every year.[118] Furthermore, the EPA data on the amount of 2,4,5-T used in the areas was "very far from reality" according to James Witt, Professor of Agricultural Chemistry at Oregon State University. Witt also points out that the EPA failed to take into account different practices for hospitalization in the three areas for women who were likely to have miscarriages. The EPA says the rate was 70% for the rural study area and assumed that it was the same for the rural control area. No attempt was made to check the value. The rate for the urban control area was 30%. Witt claims that he is not alone in finding fault with the EPA's report. In preparing his own critique of the EPA's findings Witt examined 18 other reviews of the Oregon study; all were critical and disagreed with the EPA's conclusions that there was a link between 2,4,5-T spraying and the incidence of miscarriage.[119]

This resounding "thumbs down" for the EPA study was noted by Federal Court Judge James Harvey on 12 April 1979 when he refused to grant a request from a coalition of chemical companies including Dow Chemical that the suspension on 2,4,5-T be lifted. Harvey described Dow's criticisms of the EPA study as "speculative opinion" but admitted that the criticisms from industry raised serious doubts about the findings. In refusing the companies' request, Harvey claimed to have arrived at his decision "with great reluctance" and confessed that the court "would not in its judgment have ordered the emergency suspension." He added that it was only because the court is in large part proscribed from substituting its judgment for that of the EPA that the suspension was upheld.[120]

Judge Harvey's comments have in turn been noted by the EPA. In a rearguard action, the Agency is revising part of the Oregon investigation to compute more carefully the ratio of miscarriages to live births in both the study and control areas.[120] Not everyone in the EPA is convinced that 2,4,5-T poses a health risk. On 28 September 1979 the Agency's scientific subcommittee ruled that as far as 2,4,5-T use for weed control on rice and rangeland was concerned, there was no human health hazard from spray drift. The subcommittee was still not convinced that there was sufficient evidence to indict 2,4,5-T but it recommended a further lowering of the dioxin content of the herbicide.[121,122]

There has been a perceptible shift in the Agency's strategy on 2,4,5-T to be seen in the preparation for its hearings which began in March 1980. The Agency is now putting less emphasis on the herbicide's teratogenic risk and more on its cancer threat. As dioxin has been shown to cause cancer in animals and must, therefore, be suspected as a potential human carcinogen, the EPA is now focusing on the potential carcinogenic risk from dioxin in 2,4,5-T.[123]

The chemical companies for their part have attempted to dismiss the EPA's fears. They will refer to previous incidents where 2,4,5-T was claimed to have caused animal deaths or illness in humans but which on closer scrutiny were refuted. One such case involved reindeer deaths in Swedish Lapland which were initially attributed to 2,4,5-T spraying. However, veterinarians investigating the allegations said the deaths were not caused by the herbicide, but by starvation. An alleged increase in human illnesses in Globe, Arizona, was also attributed to 2,4,5-T. The claim was dismissed after close investigation by local doctors practicing in the area.[124]

More recent claims about the dangers of 2,4,5-T have been made, only to be dismissed after further investigation. In 1977 the New Zealand Department of Health reviewed the apparent abnormally high incidence of neural tube defects (including spina bifida) in humans in three areas of North Island; the high incidence was attributed to 2,4,5-T spraying. The Department's investigation cleared the herbicide. It concluded that in two of the three areas, the birth defects were considered to be a chance occurrence, and in the third area, if there was a common causal factor it was not 2,4,5-T.[125] But a more recent publication has noted that there is an association between spraying of 2,4,5-T and the incidence of the congenital malformation talipes (clubfoot) in New Zealand. It has yet to be proved, however, that 2,4,5-T is the cause of the problem.[126]

There are similarly conflicting reports on the effects of 2,4,5-T spraying in Australia. Two recent investigations in Australia have cleared 2,4,5-T as the cause of birth defects. The first study by the Consultative Council of the Victoria Minister of Health concluded that a high incidence of birth defects in the Yarram district of Victoria was not caused by exposure to 2,4,5-T or 2,4-D.[127] A second investigation, by the Australian National Health and Medical Research Council, ruled out any link between birth defects in the State of Queensland and the burning of sugar cane stumps previously sprayed with 2,4,5-T (incineration of 2,4,5-T in the field will produce small quantities of dioxin).[128]

The first of these two investigations has been challenged, however. According to Drs. Peter Hall and Ben Selinger of the Australian National University the study did not examine all the relevant Victoria data and also makes a serious error in its statistical evaluation. Hall and Selinger claim that the rate of birth defects has indeed risen in the Yarram district although they are cautious not to impute the increase to 2,4,5-T usage.[129]

These same authors have also recently alleged that some stocks of Agent Orange left over from Vietnam may not have been disposed of by incineration. There is evidence, they claim, that some 312,000 lbs of the herbicide was imported into Australia in the late 1960s and early 1970s.[130] Hall and Selinger are concerned about the effect that the use of Agent Orange may have on the current multimillion dollar epidemiological study on the effects of herbicides on Australian veterans of the Vietnam War. The study will compare veterans exposed to defoliants with others who were not exposed and who did not serve in Vietnam. By implication, Hall and Selinger suggested that there may not be a true "unexposed" population for study in Australia if Agent Orange had been widely used. The authors called on the Australian Government to investigate the matter.

Two other authors have claimed that there is a definite relationship between 2,4,5-T spraying and neural tube defects in the Australian State of New South Wales. Analyzing the figures for 2,4,5-T usage, the authors Barbara Field and Charles Kerr claim that the incidence of such defects, including spina bifida and anencephaly, can be directly related to the previous years spraying pattern.[131]

A study of birth deformities in Hungary could find no such association, however. The incidence of stillbirths, spina bifida, and anencephaly have declined in Hungary in the period 1970–1976, while the use of 2,4,5-T has increased almost 24-fold, from 28 tonnes to 660 tonnes. The incidence of other defects such as cleft palate, cleft lip, and cystic kidney disease have remained relatively stable in this period.[132] Although more carefully case-controlled studies may show a different pattern the fact that the incidence of birth defects for Hungary as a whole bears no relationship to the increasing use of 2,4,5-T in the country has been described as "reassuring."[132]

Other reassuring reviews were published in 1978 and March 1979 by the West German Federal Health Office[133] and the U.K. Ministry of Agriculture, Fisheries and Food (MAFF).[134] Both concluded that there was no reason to restrict the use of 2,4,5-T products. Neither review was convinced that 2,4,5-T, if used for the purpose for which it was intended, presented a health risk. MAFF's verdict has been challenged by several trade unions in Britain. The union campaign led by the National Union of Agricultural and Allied Workers has resulted in the 11-million-strong Trades Union Congress calling for a ban on 2,4,5-T.[135] The herbicide has been assessed for the ninth time by MAFF's Advisory Committee on Pesticides (PAC). When it published its eighth report in 1979 the Advisory Committee estimated annual U.K. usage of 2,4,5-T at only 3 tonnes. In 1980 the Committee was embarrassed to admit that it had been out by a factor of 20 in its estimate; figures for grassland usage of the herbicide had not been taken into account. The annual tonnage of 2,4,5-T used in Britain is now put at 58 tonnes.[136]

When it finally published its latest review in December 1980, the PAC did not endorse the call for a ban on 2,4,5-T.[137] It called for tighter controls on the use of the herbicide, and welcomed the new U.K. limit of dioxin in 2,4,5-T (now less than 0.01 ppm[138,139]) and suggested that an epidemiology study might be conducted to investigate the claim that Swedish forestry workers exposed to 2,4,5-T have six times the normal incidence of soft tissue cancer.[140]

The Swedish study, by Dr. Lennart Hardell of the University of Umea's Department of Oncology, is a retrospective survey of men employed in the lumber industry in Sweden. Concerned about an apparently high incidence of soft tissue sarcomas in workers referred to his clinic, Hardell attempted to trace a causal factor. Workers and their families were asked about past exposure to chemicals, with no reference being made to 2,4,5-T. Exposure to this herbicide and to chlorophenols—and with them to dioxin—was the most frequent factor among those affected, and Hardell says this exposure could be responsible for the high incidence of cancer.

Hardell's findings have been accepted by the Swedish Medical Authorities but with some reservations. According to one of the authorities' reviewers, Professor Sune Larsson of Statens Naturvardsverk, Fack, the main reservation concerns the accuracy of reporting exposure to herbicides.[141] The herbicide 2,4,5-T has also been a subject of heated debate in Sweden and therefore much in the public eye. For this reason Larsson has some doubts that Hardell obtained unbiased information when assessing herbicide exposure. And Larsson points out that, had Hardell's information been wrong on just two of his 27 subjects, 2,4,5-T could not have been implicated as the cause of the soft tissue sarcomas. A report in 1980 on brushwood herbicides by a working group of Finland's National Board of Health expressed similar reservations.[142]

Further studies will be necessary to confirm or refute Hardell's findings. He has carried out a second investigation which implicates exposure to phenoxy acids (of which 2,4,5-T is one) and to chlorophenols as a possible cause of the cancer malignant lymphoma.[143]

Britain's Pesticides Advisory Committee did not consider Hardell's evidence sufficiently damaging to warrant a ban being placed on the herbicide. This opinion did not satisfy the trades unions, however. They have called the 1980 PAC report a "whitewash" and called for the committee to be "wound up," and replaced by a review body on which the unions and industry would be represented and where both would have a say on the safety, or otherwise, of herbicides.[144]

The NUAAW, which has been in the forefront of the campaign to have 2,4,5-T banned in the United Kingdom, argued that the evidence from Sweden that dioxin might cause soft tissue cancers has been supported by more recent studies in the United States.[145,146] Workers exposed to dioxin 12-32 years previously at the Dow and Monsanto chemical companies have an incidence of soft tissue cancer 41 times that of the general population. The concentrations of dioxin to which the U.S. workers were exposed were many times higher than those which occurred in Sweden or are likely to occur in the United Kingdom. Not unexpectedly, the union's campaign against 2,4,5-T in Britain has had an effect on sales. Usage of the herbicide fell from 58 tonnes in 1979 to 38 tonnes in 1980.[147]

Sales of 2,4,5-T have also slumped in the United States following the partial ban on the herbicide imposed by the Environmental Protection Agency. It is the Agency's stated view that 2,4,5-T with its dioxin contaminant poses a significant health hazard, and the recent evidence on soft tissue cancer from Sweden and the United States merely reinforces its position.

Ten years after it first took up the cudgels and imposed some restrictions on the herbicide, the EPA is now convinced that 2,4,5-T should be banned. The Dow Chemical Company argues to the contrary. According to a Dow spokesman, Mr. Gary Jones, the company, in opposing the call for a ban, sees itself as defending an attack on the chemical industry by environmentalists who "want to go back to windmills."[76] As 2,4,5-T has been in use for nearly 40 years, Dow's view, says Jones, is that "with all the confidence we have on the safety of 2,4,5-T, we can't say it is safe, then we can't prove the safety of aspirin."

A verdict in favor of a ban could well result in Dow facing a series of expensive compensation suits from individuals claiming to have health problems which arose following exposure to the herbicide. A ruling against the ban would mean a serious loss of face for the EPA. With the stakes so high it came as no surprise to many that the court hearing on 2,4,5-T was adjourned early in March 1981 to allow the lawyers representing both parties to see if a compromise formula might be reached. The adjournment was extended on three separate occasions to allow negotiations to continue, and the court hearings were scheduled to begin again in July 1981. According to officials within the EPA there was pressure from the new Reagan government to reach some kind of agreement relating to some restrictions on the use of 2,4,5-T and a lowering, or complete removal, of the dioxin contaminant in the herbicide. Dow, it was thought, would fare better than the EPA in any such deal. In the event of a compromise not being possible, both sides would take up their old adversary positions.

The announcement of the discussions between the EPA and Dow stunned many environmental groups in the United States. Lawyers representing Vietnam War veterans claiming compensation for health problems which they claim have been caused by Agent Orange were not happy about the announcement either. In any event a solution to the Dow/EPA conflict is unlikely to have much impact on the veterans' case.[105] The veterans have argued that their problems were caused by herbicides being contaminated with high concentrations of dioxin. If Dow agrees to lower the dioxin levels in its 2,4,5-T formulations this will be taken by the veterans as proof that the company accepts that dioxin is a health hazard. Some argue that it may even strengthen the veterans' case.

The sums of money involved in the compensation suits are vast. It is ironic that a herbicide which was developed as another weapon in the U.S. arsenal and which was a profitable revenue earner for chemical companies could end up as an expensive liability for government and companies alike. The fruit of their collective endeavors may yet turn out to be sour.

References

1. Perez, D. D. and Gerrol, D. M. *The Observer*, 24 January 1969.
2. National Academy of Sciences, The effects of herbicides in South Vietnam. Part A, summary and conclusions, Washington, D.C. 1974, p. S-3.
3. National Academy of Sciences, The effects of herbicides in South Vietnam. Part A, summary and conclusions, Washington, D.C. 1974, p. S-6.
4. Westing, A. Ecological consequences of the second Indo-China War, Stockholm International Peace Research Institute (1976), p. 28.
5. Background report on herbicide operations in South Vietnam, Department of Defense (April 1971).
6. Evaluation of carcinogenic, teratogenic and mutagenic activities of selected pesticides and industrial chemicals, Vol. II, teratogenic study in mice and rats, Bionetics Research Laboratories Inc. National Technical Information Service, Doc. No. PB 223-160 (1969).
7. Whiteside, T. *The Withering Rain: America's Herbicidal Folly*, E. P. Dutton and Co., New York (1971).
8. Irish, K. P., Darrow, R. A., and Minarik, C. E. Information manuals for vegetation control in South East Asia, Department of the Army, Fort Detrick, Maryland. Miscellaneous Publication 33, 71 pp. (1969).
9. Assessment of ecological effects of extensive or repeated use of herbicides, Midwest Research Institute Contract No. DAHC15-68-0119 (1967).
10. Anonymous, *Potentialities of weapons of war during the next ten years*, prepared for Joint Technical Warfare Committee (12 November 1945), and reported by Norton-Taylor, Richard, *The Guardian* (5 May 1981).
11. Osborne, D. J. Defoliation and defoliants, *Nature (London)* **219**, 564-67 (1968).
12. Clutterbruck, R. *The Long, Long War: The Emergency in Malaya 1948-60*, Cassell, London (1967), p. 160.
13. Hamner, C. L., and Tukey, H. B. The herbicidal action of 2,4-dichlorophenoxyacetic and 2,4,5-trichlorophenoxyacetic acid on bindweed, *Science* **100**, 154-55 (1944).
14. Fryer, J. D., Johns, D. L., and Yeo, D. The effect of a chemical defoliant on an isolated tsetse fly community and its vegetation, *The Bulletin of Entomological Research* **48**, 359-373 (1957).
15. Carey, W. C. Engineer uses of chemical herbicides, *Military Engineer* **44**, 174-177 (1952).
16. Hillsman, R. *To Move a Nation*, Doubleday and Co., New York (1967).
17. Bovey, R. W., and Young, A. L. *The Science of 2,4,5-T and Associated Phenoxy Herbicides*, John Wiley, New York (1980), p. 379.
18. Young, A. L., Calcagni, J. A., Tahlken, C. E., and Tremblay, J. W. The toxicology, environmental fate and human risk of herbicide orange and its associated dioxin, USAF Occupational and Environmental Health Laboratory, Aerospace Medical Division (AFSC), Brooks Air Force Base, Texas 78235 (October 1978), I-9.
19. Westing, A. Ecological consequences of the Second Indo-China War, Stockholm International Peace Research Institute (1976), p. 27.

20. A technology assessment of the Vietnam defoliant matter, a report of the Sub-committee on Science, Research and Development, of the Committee on Science and Astronautics, U.S. House of Representatives (8 August 1969), p. 28.
21. Galston, A. Personal interview, 11 January 1979.
22. Petition to President Lyndon B. Johnson, 14 February 1967.
23. A technology assessment of the Vietnam defoliant matter (8 August 1969), p. 22.
24. Mayer, J. Starvation as a weapon. Herbicides in Vietnam, *Scientist and Citizen* **9**, 115–121 (1967).
25. Galston, A. W. Changing the environment. Herbicides in Vietnam II, *Scientist and Citizen* **9**, 122–129 (1967).
26. A technology assessment of the Vietnam defoliant matter (8 August 1969), p. 32.
27. A technology assessment of the Vietnam defoliant matter (8 August 1969), pp. 32–33.
28. A technology assessment of the Vietnam defoliant matter (8 August 1969), p. 34.
29. A technology assessment of the Vietnam defoliant matter (8 August 1969), pp. 38–40.
30. A technology assessment of the Vietnam defoliant matter (8 August 1969), p. 44.
31. A technology assessment of the Vietnam defoliant matter (8 August 1969), p. 46.
32. A technology assessment of the Vietnam defoliant matter (8 August 1969), p. 47.
33. A technology assessment of the Vietnam defoliant matter (8 August 1969), pp. 47–48.
34. Tschirley, F. An assessment of ecologic consequences of the defoliation programme in Vietnam, a report prepared at the request of the U.S. Department of State (12 April 1968).
35. Tschirley, F. Defoliation in Vietnam, *Science* **163**, 779–786 (1969).
36. Flamm, B. R. A partial evaluation of herbicidal effects to natural forest stands in Tay Ninh province, USAID/ADDP (15 April 1968).
37. A technology assessment of the Vietnam defoliant matter (8 August 1969), p. 48.
38. A technology assessment of the Vietnam defoliant matter (8 August 1969), p. 49.
39. A technology assessment of the Vietnam defoliant matter (8 August 1969), p. 57.
40. Society for social responsibility in science, Ecological studies—ICR (27 February 1969).
41. Pfeiffer, E. Personal interview, 13 January 1969.
42. Department of State telegram, Ref. Saigon 4979, 19 March 1969.
43. Department of State telegram. From US Ambassador Ellsworth Bunker, American Embassy Saigon, March 1969.
44. Pfeiffer, E. W., and Orians, G. H. The Military uses of herbicides in Vietnam, in *Harvest of Death: Chemical Warfare in Vietnam and Cambodia*, ed. J. B. Nielands, G. H. Orians, E. W. Pfeiffer, A. Vennema, and A. H. Westing, Free Press, New York, Collier-Macmillan Limited, London (1972), pp. 117–176.
45. Sullivan, W. *The New York Times*, 4 April 1969.
46. Mission to Vietnam, Scientific Research Part 1 (9 June 1969), Part 2 (23 June 1969).
47. Report of the Secretary's Commission on Pesticides and their Relationship to Environmental Health, U.S. Department of Health, Education and Welfare (December 1969).
48. Whiteside, T. *The Withering Rain: America's Herbicidal Folly*. E. P. Dutton and Co., New York (1971), p. 49.
49. Bovey, R. W., and Young, A. L. *The Science of 2,4,5-T and Associated Phenoxy Herbicides*, John Wiley, New York (1980), p. 380.
50. Johnson, A. Personal communication, 16 January 1979.
51. DuBridge Statement, Executive Office of the President, Office of Science and Technology, 29 October 1969.
52. Whiteside, T. *The Withering Rain: America's Herbicide Folly*, E. P. Dutton and Co., New York (1971), p. 41.
53. Whiteside, T. *The Withering Rain: America's Herbicide Folly*, E. P. Dutton and Co., New York (1971), p. 50.
54. *Science* **167**, 36 (1970).

55. Senate Congressional Record, S13226-S3233 (3 March 1972).
56. Young, A. L., Calcagni, J. A., Thalken, C. E., and Tremblay, J. W. The toxicology, environmental fate and human risk of herbicide orange and its associated dioxin, USAF Occupational and Environmental Health Laboratory, Aerospace Medical Division (AFSC), Brooks Air Force Base, Texas 78235 (October 1978), p. I-3.
57. Disposition of orange herbicide by incineration, final environment statement, Department of the Air Force, Washington, D.C. (1974), 737 pp.
58. Anonymous, *New York Times*, 24 October 1970.
59. Meselson, M. Personal interview, 16 January 1979.
60. Cutting, R. T., Phuoc, T. H., Ballow, J. M., Berenson, M. W., and Evans, C. H. *Congenital Malformations*, Hydatidiformmoles and Stillbirths in the Republic of Vietnam 1960-69, U.S. Government Printing Office, Washington, D.C. (1970).
61. Cohn, V. *San Francisco Chronicle*, 31 December 1970.
62. Bevan, W. Letter to E. W. Pfeiffer, 20 January 1971.
63. Young, A. L., Calcagni, J. A., Thalken, C. E., and Tremblay, J. W. The toxicology, environmental fate and human risk of herbicide orange and its associated dioxin, USAF Occupational and Environmental Health Laboratory, Aerospace Medical Division (AFSC), Brooks Air Force Base, Texas 78235 (October 1978), p. I-3.
64. Shapley, D. Herbicides: Agent Orange stockpile may go to the South Americans, *Science* **180**, 43-45 (1973).
65. Mintz, M. *Washington Post*, 25 September 1974.
66. Anonymous. Environmental impact analysis, amendment to final environmental impact statement on disposition of herbicide Orange by incineration, Department of the Air Force (SAF/ILE) Washington, D.C. (October 1976).
67. Anonymous. Dioxin tainted ship, *Chemical Week*, 7 June 1978.
68. Boffey, P. M. Academy may study defoliation, *Science* **170**, 43 (1970).
69. Bovey, R. W., and Young, A. L. *The Science of 2,4,5-T and Associated Phenoxy Herbicides*, John Wiley, New York (1980), p. 5.
70. National Academy of Sciences. The effects of herbicides in South Vietnam. Part A. Summary and Conclusions, Washington, D.C. (1974), p. VII-9.
71. Courtney, K. D., Gaylor, D. W., Hogan, M. D., Falk, H. L., Bates, R. R., and Mitchell, I. Teratogenic evaluation of 2,4,5-T, *Science* **168**, 864-866 (1970).
72. Courtney, K. D., and Moore, J. A. Teratology studies with 2,4,5-trichlorophenoxyacetic acid and 2,3,7,8-tetrachlorodibenzo-p-dioxin, *Toxicology and Applied Pharmacology* **20**, 396-403 (1971).
73. Sparschu, G. L., Dunn, F. L., and Rowe, V. K. Study of the teratogenicity of 2,3,7,8-tetrachlorodibenzo-p-dioxin in the rat, *Food and Cosmetic Toxicology* **9**, 405-412 (1971).
74. Report on 2,4,5-T, a report of the Panel on Herbicides of the President's Science Advisory Committee, Executive Office of the President, Office of Science and Technology (March 1971).
75. Report of the Advisory Committee on 2,4,5-T to the Administrator of the Environmental Protection Agency (7 May 1971).
76. Hay, A. Dioxin: the 10 year battle that began with Agent Orange, *Nature (London)* **278**, 108-109 (1979).
77. Respondents prehearing brief on the risks associated with the registered uses of 2,4,5-T and Silvex, U.S. Environmental Protection Agency, FIFRA Docket Nos. 415 *et al.*
78. Statement of Prime Minister Olaf Palme in Plenary Meeting June 6, 1972, Swedish delegation to the U.N. Conference on the Human Environment.
79. Westing, A. Personal interview, 16 January 1979.
80. Hill, G. *New York Times*, 8 June 1972.
81. National Academy of Sciences. The effects of herbicides in South Vietnam. Part A. Summary and Conclusions, Washington, D.C. (1974), pp. XII-XIV.

82. Norman, C. Academy reports on Vietnam's herbicide damage, *Nature (London)* **248**, 186–188 (1974).
83. Hay, A. U.S. Defense Department destroys evidence of Vietnam devastation, *Nature (London)* **279**, 662 (1979).
84. National Academy of Sciences, The Effects of Herbicides in South Vietnam. Part A. Summary and Conclusions, Washington, D.C. (1974), p. X.
85. Baughman, R., and Meselson, M. An analytical method for detecting TCDD (Dioxin): Levels of TCDD in samples from Vietnam, *Environmental Health Perspectives* **5**, 27–36 (1973).
86. National Academy of Sciences, The effects of herbicides in South Vietnam. Part A. Summary and Conclusions, Washington, D.C. (1974), p. XI.
87. Tung, T. T., Anh, T. K., Tuyon, B. Q., Tra, D. X., and Huyan, N. X. Clinical effects on the civilian population as a result of the massive and continuous use of defoliants (preliminary study); *Vietnamese Studies* **29**, 53–81 (1971).
88. Hay, A. Vietnam's dioxin problem, *Nature (London)* **271**, 597–598 (1978).
89. Tung, T. T. Pathologie humaine et animale de la dioxine, *La Revue de Medicine* **14**, 653–657 (1977).
90. Wade, N. Viets and Vets fear herbicide health effects, *Science* **204**, 817 (1979).
91. Tung, T. T., Lang, T. D., and Van, D. D. Le probleme des effets mutagenes sur la premiere generation apres l'exposition aux herbicides (unpublished) (1979).
92. Clarke Fraser, F. In *Handbook of Teratology*, ed. J. G. Wilson and F. Clarke Fraser, Plenum Press, New York (1977), p. 75.
93. McClaren, A. Director MCR Mammalian Development Unit, London, Personal communication, 27 August 1980.
94. Lamb IV, J. C., Moore, J. A., and Marks, T. A. Evaluation of 2,4-dichlorophenoxyacetic acid (2,4-D), 2,4,5-trichlorophenoxyacetic acid (2,4,5-T) and 2,3,7,8-tetrachlorodibenzo-*p*-dioxin (TCDD) toxicity in C57BL/6 mice. Reproduction and fertility in treated male mice and evaluation of congenital malformations in their offspring, National Toxicology Program NTP-80-44 (1980).
95. Bogen, G. Symptoms in Vietnam veterans exposed to Agent Orange, *Journal of the American Medical Association* **242**, 2391 (1979).
96. Holden, C. Agent Orange furor continues to build, *Science* **205**, 770–772 (1979).
97. Rawls, R. L. Dow finds support, doubt for dioxin ideas, *Chemical and Engineering News* **57**, 23–29 (1979).
98. Severo, R. *The New York Times*, 27 May 1979.
99. Severo, R. *The New York Times*, 28 May 1979.
100. Wolfe, W. H. Human health effects following exposure to the phenoxyherbicides and TCDD, *Proceedings of the Educational Conference on Herbicide Orange*, U.S. Veterans Administration, Silver Spring, Maryland (28–30 May 1980).
101. Bovey, R. W., and Young, A. L. *The Science of 2,4,5-T and Associated Phenoxy Herbicides*, John Wiley, New York (1980), pp. 400–401.
102. Binder, D. *The New York Times*, 29 May 1979.
103. Wolfe, W. H. Diagnostic indicators of phenoxyherbicide/dioxin exposure, Proceedings of Educational Conference on Herbicide Orange, U.S. Veterans Administration, Silver Spring, Maryland (28–30 May 1980).
104. Yannacone, Jr. V. J., Personal letter, 11 July 1980.
105. Kavenagh, D. K. of Yannacone and Yannacone, personal communication (8 May 1981).
106. Young, A. L. Use of herbicides in South Vietnam 1961–1971, in Proceedings of Educational Conference on Herbicide Orange, U.S. Veterans Administration, Silver Spring, Maryland (28–30 May 1980).
107. Anonymous. Defoliant companies can be sued, *Nature (London)* **282**, 434 (1979).
108. Kavenagh, D. K. of Yannacone and Yannacone, personal communication, 2 June 1980.

109. Anonymous. *The Daily Telegraph*, March 1980.
110. Anonymous. *The Guardian*, 8 August 1980.
111. Tybor, J. R. Agent Orange: a red alert, *The National Law Journal* (13 October 1980).
112. Anonymous. Veterans lose agent orange case, *New Scientist* **88**, 693 (1980).
113. Jackson, H. *The Guardian*, 31 December 1980.
114. Omang, J. *The Washington Post*, 15 March 1980.
115. Cookson, C. Emergency ban on 2,4,5-T herbicide in U.S., *Nature (London)* **278**, 108–109 (1979).
116. Report of assessment of a field investigation of six year spontaneous abortion rates in three Oregon areas in relation to forest 2,4,5-T spray practices, Environmental Protection Agency (28 February 1979).
117. Hay, A. Critics challenge data that led to 2,4,5-T ban, *Nature (London)* **279**, 3 (1979).
118. Wagner, S. L., Witt, J. M., Norris, L. A., Higgins, J. E., Aresti, A., and Ortiz, M. A scientific critique of the EPA Alsea II study and report, Environmental Health Sciences Center, Oregon State University (25 October 1979).
119. Witt, J. M. A discussion of the suspension of 2,4,5-T and the EPA Alsea II study.
120. Smith, R. J. Court reluctantly upholds EPA on 2,4,5-T suspension, *Science* **204**, 602 (1979).
121. Anonymous. Report of the FIFRA Scientific Advisory Panel on proposed section 6(b)(2) notices for 2,4,5-T and Silvex, Environmental Protection Agency (27 September 1979).
122. Hay, A. EPA panel recommends postponement of 2,4,5-T hearings. *Nature (London)*, **282**, 124 (1979).
123. Anonymous, Dioxin and 2,4,5-T: what are the risks? *Nature (London)* **284**, 111 (1980).
124. Bovey, A. W., and Young, A. L. *The Science of 2,4,5-T and Associated Phenoxy Herbicides*, John Wiley, New York (1980), pp. 11–12.
125. Anonymous. 2,4,5-T and human birth defects, New Zealand Department of Health (June 1977).
126. Hanify, J. A., Metcalf, P., Nobbs, C. L., and Worsley, K. J. Aerial spraying of 2,4,5-T and human birth malformations: an epidemiological investigation, *Science* **212**, 349–351 (1981).
127. Anonymous. Consultative Council on Congenital Abnormalities in the Yarram District, Victoria, Australia, Report, Department of Primary Industry, Canberra (1978).
128. Press statement. Australian National Health and Medical Research Council, 16 June 1978.
129. Hall, P., and Selinger, B. Australian infant mortality from congenital abnormalities of the central nervous system: a significant increase in time. Is there a chemical cause? *Chemistry in Australia* (in press).
130. Hall, P., Selinger, B., Field, B., and Kerr, C. Antipodean 2,4,5-T, *Nature (London)* **290**, 8 (1981).
131. Field, B., and Kerr, C. Herbicide use and incidence of neural-tube defects, *Lancet* **I**, 1341–1342 (1979).
132. Thomas, H. F. 2,4,5-T use and congenital malformation rates in Hungary, *Lancet* **II**, 214–215 (1980).
133. Anonymous. Commentary on the use of 2,4,5-T for purposes of weed control in forests, Mitteilungen aus der biologischen Bundesanstalt fur land-und Forstwirtschaft, No. 181 (1978).
134. Anonymous. Review of the safety for use in the UK of the Herbicide 2,4,5-T, Advisory Committee on Pesticides, Ministry of Agriculture, Fisheries and Food (March 1979).
135. Anonymous. *The Guardian*, 5 September 1980.
136. Hay, A. Red faces (and hot tempers) on 2,4,5-T. *Nature (London)* **286**, 97 (1980).
137. Anonymous. Further review of the safety for use in the UK of the herbicide 2,4,5-T, Ad-

visory Committee on Pesticides, Ministry of Agriculture, Fisheries and Food (December 1980).
138. Anonymous. *The Daily Mirror*, 23 May 1980.
139. Hay, A. A plea for 2,4,5-T, *Nature (London)* **287,** 569–570 (1980).
140. Hardell, L., and Sandstrom, A. Case-control study: of tissue sarcomas and exposure to phenoxyacetic acids at chlorophenols, *British Journal of Cancer* **39,** 711–717 (1979).
141. Larsson, S. Personal communication, 3 September 1980.
142. Anonymous, Brushwood herbicides, The National Board of Health Working Group Report No. 4, Helsinki (1980).
143. Hardell, L., Eriksson, M., Lenner, P., and Lundgren, E. Malignant lymphoma and exposure to chemicals, especially organic solvents, chlorophenols and phenoxy acids: a case control study. *British Journal of Cancer* **43,** 169–176 (1981).
144. Press release accompanying "Pray before you Spray," U.K. National Union of Agricultural and Allied Workers (27 April 1981).
145. Honchar, P. A., and Halperin, W. E. 2,4,5-T, Trichlorophenol and soft tissue sarcoma, *Lancet* **I,** 268–269 (1981).
146. Cook, R. R. Dioxin, chloracne and soft tissue sarcoma, *Lancet* **I,** 618–619 (1981).
147. Erlichman, J. *The Guardian*, 20 May 1981.

8

2,4,5-T in Cambodia and Laos

Cambodia

Discussion of the use of herbicides in Indochina would be incomplete without reference to a defoliation mission carried out over Cambodia in the Spring of 1969.[1-4] The timing of the incident is significant because at the time the Fishook area of Cambodia was being sprayed it is also alleged to have been bombed. Attacks on Cambodia, a neutral state, were violations of international law and hence illegal. The illegal bombing of the country by B-52 bombers of the U.S. Air Force is said to have occurred in April 1969.[5,6]

One hundred and seventy three thousand acres in Cambodia were affected by herbicides between mid-April and mid-May 1969. Dating cannot be more precise as no agency has ever claimed responsibility for the mission and hence no records are available for checking. However, on 19 May 1969, a letter of protest from the Cambodian Minister of Foreign Affairs was sent to the United States drawing attention to the damage caused and seeking reparation.[1] Four days later at a press conference in the capital Phnom Penh, Cambodia's Head of State, Prince Sihanouk, spoke of an "unprecedented catastrophe" in the region of Kampong Cham.[1] The Prince's denunciation was followed by action at the Security Council of the United Nations. On 26 May Cambodia's permanent representative to the U.N., Huot Sambath, informed the Security Council that "From 18 April to 2 May 1969, aircraft of the United States–South Vietnamese air forces continually scattered defoliants every two days over a vast area of Cambodia stretching along the Khmer–Vietnamese frontier and extending to a depth of 20 kilometers from the frontier, in the districts of Krek and Mimot, province of Kompong Cham."[7] The letter, reporting damage to 7000 hectares of rubber plantations and vast areas of woodland, expressed the indignation of the Cambodian government.

In a letter of 17 June to the President of the Security Council, Sambath gave more details of the damage. The defoliation missions, he now claimed, were flown between 19 April and 12 May 1969. An area of approximately 85,000 hectares was affected and he assessed the damage at 7.6 million U.S. dollars.[8] Some months later in December 1969, a more detailed survey, at the invitation of the Cambodian government, was carried out by the botanist Professor Arthur Westing (now Dean at the School of Natural Sciences, Hampshire College, Amherst, Massachusetts), the zoologist Professor E. W. Pfeiffer of the University of Montana, and two French colleagues, a biophysicist Dr. Jean Lavorel and lawyer M. Leon Matarasso.[1] In a report of the survey, Westing estimated the damage at 12.2 million U.S. dollars; 173,000 acres had been sprayed, said Westing, of which about 24,700 were seriously affected.[1]

Agent Orange (the 1:1 mixture of 2,4-D and 2,4,5-T) and Agent White (2,4-D plus picloram) were the herbicides in general use by U.S. forces in Southeast Asia in 1969. Surveillance of affected areas in Kompong Cham province suggested that Agent Orange was the principal herbicide responsible.[1] A U.S. Department of State investigative team stated that much of the damage in the immediate vicinity of the border had been caused by spray drifting across from herbicide operations in Vietnam.[2] Where there was evidence of more serious damage which could not be explained by spray drift and weather conditions at the time, it was suggested that herbicides had been directly applied to the area and that one or two UC-123 planes (the Air Force planes equipped for herbicide operations) could have been responsible. Westing disagrees with this estimate for several reasons. He points out that the State Department survey was in July 1969 and only two months after the operation. Westing's own survey was five months later, in December 1969, and he claims it was able to present a clearer picture of the extent of the damage. This was consistent with chemicals of far greater concentrations being used than the State Department officials had estimated.[2] Westing's figures led him to conclude that the number of aircraft involved in direct flights over Cambodia was at least seven, perhaps more.[1] Cambodian reports referred to seven or eight separate missions without specifying the number of aircraft involved.[1]

The type of vegetation affected and the extent of the damage caused are well documented in the accounts of both Westing's survey and that of the State Department officials. Only Westing gives any details of the effects of spraying on the health of the population and his report is anecdotal, based as it is on interviews with the local inhabitants. However, the answers to his questions were consistent and corroborative. Temporary diarrhea and vomiting had been widespread at the time of spraying, infants being the worst affected. To avoid consumption of contaminated drinking water, fresh supplies were ferried in for a short time. Where people relied on wells as a source of drinking water there were no reports of digestive problems.

No assessment could be made of any change in the incidence of congenitally malformed children or spontaneous abortions directly attributable to the spraying. Although Westing and his colleagues interviewed people eight months after the herbicide operation—a time when any increase in the incidence would have been detectable—his population survey was of a limited nature. Thus, says Westing, "Our negative findings are, of course, not definitive, owing to the very small sample involved."[1]

An estimated 30,000 people are thought to have been living in the area of Kompong Cham province which was sprayed. Of the 173,000 acres affected, some 38,300 acres were rubber plantations, representing about 30% of Cambodia's rubber production. The compensation claim lodged by the Cambodian Government was in large part designed to cover the losses incurred by the several major French companies with holdings in the area and to a lesser extent the many hundreds of privately owned plantations.[1]

In processing the claim, the United States made it abundantly clear that the Government, having failed in its attempt to find out who was responsible for the spraying, was processing the claim; but neither accepted nor rejected responsibility for the operation.[9] U.S. Senator Daniel K. Inouye of Hawaii, a member of the Senate Select Committee on Intelligence, was asked to enquire about the current position regarding the claim. Inouye says the position is as it was before and the claim has not been resolved.[10] Having been filed by a government which was subsequently overthrown in a coup in March 1970,[4] the claim entered a sort of limbo for there were no longer any legal claimants. According to the Senator there was some indication that the government of France was going to bring the claims on behalf of certain companies, but for some reason declined to do so.[10]

The damage caused by herbicides in this region of Cambodia is in many respects an academic issue, in view of the fact that the area was subjected to repeated heavy saturation bombing and ground battles when U.S. forces invaded Cambodia on 30 April 1970. The region, by all accounts, was largely devastated.[1] According to some reports, however, some of this devastation occurred a year earlier when the region was subjected to an illegal attack by U.S. B-52 bombers.[5,6]

The allegations of illegal bombing were first made by journalist William Shawcross in his book *Sideshow*.[5] Shawcross rests his case on classified information obtained under the Freedom of Information Act from various U.S. government agencies including the Department of Defense, the State Department, Central Intelligence Agency, and National Security Council. Quoting from source material obtained, Shawcross builds up a case to support the thesis of illegal attack by B-52 bombers of the Fishook area. The operation was so secret that official records were falsified to keep the operation hidden. Secrecy was of the essence for, in carrying out these operations, the United States was deliberately violating the borders of a neutral state. According to Shawcross, U.S. intelligence sources mentioned the presence of large numbers of North Vietnamese and Vietcong soldiers in base camps in the Fishook area. The camps were reportedly used by the Vietnamese for foray operations across the border to attack South Vietnamese and U.S. forces. (According to some, the existence of North Vietnamese base camps in the Fishook region is in doubt. Neither Westing[1] nor Sihanouk[4] refer to their existence). It was important, therefore, that they be eliminated. But to do so, as Shawcross notes, required an operation which could not be traceable. General Earle G. Wheeler, Chairman of the Joint Chiefs of Staff, summed up the situation admirably in a briefing paper to cope with press enquiries about the bombing. Wheeler noted that "Should the press persist in its inquiries (about the bombing) or in the event of a Cambodian protest concerning U.S. strikes in Cambodia, U.S. spokesmen will neither confirm nor deny reports of attacks on Cambodia, but state it will be investigated. After delivering a reply to any Cambodian protest, Washington will inform the press that we have apologized and offered compensation."[5]

The essence of the action suggested by the General—neither confirmation nor denial, the promise of investigation, the proffered compensation—offers a suitable summary of the events which followed the herbicide spraying missions. It is still puzzling that protests which followed the herbicide missions made no reference to any B-52 bombardment. But perhaps the strangest aspect of the affair is the absence of any reference to bomb damage in the three reports describing herbicide damage in the Fishook area of Cambodia.[1-3] These very points have been raised by the author of one of the reports, Arthur Westing, in a letter to Shawcross.[11] On the question of the evidence to support bomb damage, Westing wrote that he ". . . did not see the results of any B-52 strikes in the region (despite my complete familiarity with them from Viet Nam) nor did I hear any mention of them. Moreover, the local residents were going about their business without any apparent preparation for or fear of air strikes even closer to the border." Westing continues "What is so puzzling about the B-52 strikes you describe is that I believe we would have seen them—or even been taken to them—or at least heard about them."[11] According to Westing, Cambodian officials were furious about any violations of Cambodian neutrality and took great pains to point them out to visitors. Westing is quite certain, therefore, that the Fishook area was not attacked by B-52s before December 1969.[11] Shawcross, Westing insists, is wrong on this point. Westing received no answer to his enquiries. A reminder to Shawcross about the issue went unanswered too.[12] Shawcross has said that he has noted Westing's objections, that the herbicide operation was not part of his story, and that he stood by his original allegations about the bombing missions.[13]

Shawcross is not alone in his claim that B-52 bombers struck Cambodia early in 1969. Support for his allegations come from former U.S. Secretary of State, Dr. Henry Kissinger. In his memoirs *The White House Years*, Kissinger confirms discussions about the B-52 raids with former U.S. President Richard Nixon, cabinet officials, and senior military personnel.[6] On the question of the raids themselves, Kissinger claims that "The B-52 attack took place on March 18 against North Vietnamese Base Area 353, within three miles of the Cambodian border."[6] The mission was codenamed "Breakfast." Successive missions in other areas were referred to as "Dessert," "Snack," "Lunch," "Supper," and "Dinner." The whole operation was appropriately named "Menu."[6]

"Breakfast" and "Dinner" refer to Base camps 353 and 352, respectively. Both are in the region visited by Westing in December 1969 and described by him as having been attacked with herbicides[1] (see Figure 15). The Department of Defense is more equivocal about the coincidence of the base camps and the defoliated area. In a letter to Senator Max Baucus of Montana on 5 March 1980, Principal Deputy Under Secretary of Defense Walter LaBerge, confirming that defoliation did occur in the Fishook area, claimed that his Department had been unable "To relate the alleged bombing" to the area defoliated.[14] The Senator, however, believes that the areas do coincide, and has asked the DOD to reexamine the issue in view of the "inconsistencies and contradictions" in the various accounts of what happened in the Fishook area.[15] In a later letter to the Senator, on 20 October 1980, Charles W. Hinkle, Director of the Department of Defense's Freedom of Information and Security Review, wrote that "we could find no records that indicate base areas 352 and 353 had ever been bombed prior to February 1970."[16] The DOD thus appears to confirm Westing's allegations that Shawcross is wrong about the illegal bombing of Cambodia in the spring of 1969.

Attempts have been made to find out who authorized the herbicide operations over Cambodia. Professor E. W. Pfeiffer, who accompanied Westing to Cambodia in December 1969, has been persistent in this regard, but to little or no avail. Pfeiffer's only lead was a letter addressed to him by Senator Frank Church of Idaho. Church stated that he had been told that "Air America [a then wholly owned Central Intelligence Agency subsidiary and an airline operating in Southeast Asia[17]] was responsible for the Cambodian defoliation. My source was not the State Department, but rather an individual who is in a position to know the facts in this matter."[18] The CIA has, however, denied any involvement in the affair[19]—a denial accepted by Senator Inouye.[10] As for Senator Church, he told the author that "A thorough search of the Committee on Foreign Relations files has not uncovered a public document confirming that Air America was responsible for the Cambodian defoliation."[20]

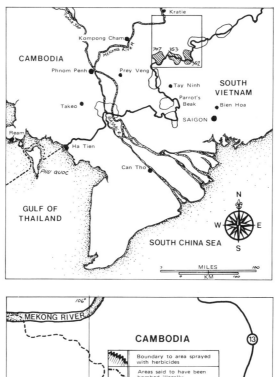
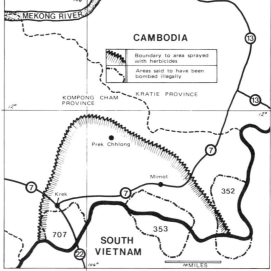

FIGURE 15. Herbicide spraying in Cambodia. The upper map shows the location of the region in Cambodia which was sprayed with herbicides. In the lower map the extent of herbicide damage is shown in more detail. Note how the region known to have been sprayed overlaps with the areas alleged to have been bombed (see text).

Everyone would agree with Defense Department spokesman Walter LaBerge's comment that "The purpose of defoliation in this area (Fishook) is still unknown."[14] Equally intriguing is the coincidence of the areas known to have been defoliated and those alleged to have been bombed. That defoliation occurred is beyond dispute, but, as the preceding section has shown, the allegations that bombing did too is questionable. The contradictions in the various stories leave many questions unanswered. Who carried out the defoliation missions and for what purpose? Why should former Secretary of State Kissinger refer to B-52 bombing operations over Cambodia when there is little evidence that they occurred, yet refrain from mentioning defoliation missions when there is ample documentation confirming that they took place? Asked to comment on these questions, Dr. Kissinger declined to do so. It seems that answers will only be forthcoming after much more intensive probing.

Laos

Secrecy was also the order of the day as far as herbicide spraying in another area of Indochina was concerned. A recent, but as yet unpublished manuscript by a U.S. Department of Defense historian William Buckingham[1] has revealed that defoliants were used in Laos from 1965 to 1969. According to Buckingham the U.S. Ambassador in Vientiane, William H. Sullivan, at first opposed the decision to extend herbicide missions to Laos. Sullivan was not so much worried about the effect of herbicides on Laos; he was far more concerned about the reaction from diplomats of nations friendly to the United States. Allegations had already been made that the United States had used chemical weapons in Laos and diplomats were sensitive about the subject. To start using herbicides to defoliate vegetation on the Ho Chi Minh trail in the country would only make matters worse, Sullivan informed the State Department in January 1965. (Regrettably, there is no additional information about the allegations that chemical weapons had been used in Laos before 1965.)

By November 1965, however, Sullivan was no longer so resolute. His position of outright opposition to spraying missions had changed. It is assumed that his change of mind was due in part to the fact that the air war had already extended into this part of Laos; the area was being heavily bombed by late 1965. Thus any further damage brought on by herbicides would not cause so much of an outcry.

It was Sullivan who selected the routes in Laos to be sprayed. Approval for the spraying missions was authorized by Secretary of State Dean Rusk and Secretary of Defense Robert McNamara on November 25, 1965. Rusk and McNamara assumed that there might be some public hostility to the missions. To try and contain this hostility, they issued a directive concerning the answering of press inquiries on the subject. Any queries were to be dealt with according to the existing guidelines which were not to report, acknowledge, or otherwise comment on U.S. air operations in Laos except to state that since May 1964 the United States had flown air reconnaissance missions over the country at the request of the Laotian authorities.

The directive was clearly effective, for no specific details about these missions ever appeared in the press. Buckingham's manuscript is the first official publication to refer to the herbicide missions over Laos. At the time of writing, Buckingham's report was still unpublished, and it is not clear when the Department of Defense will give permission for its release.

Most of the herbicide missions flown over Laos were in 1966, and of the 420,000 gallons of herbicide sprayed approximately 83% was in the form of Agent Orange. This herbicide was used mainly for defoliating dense vegetation along lines of communication. But those were not the only targets; farmland was also hit. In November 1968 and September 1969 there were two missions flown purely for crop destruction. The arsenic-based preparation Agent Blue was used on those last two occasions to kill 3900 acres of crops. According to Arthur Westing,[22] the crop land and forestry sprayed in Laos totaled some 165,000 acres or 0.2% of the land in the country. The herbicide used represents about 2% of the total amount sprayed by U.S. forces in Indochina.

From this evidence it is now clear that when herbicide spraying in Indochina is discussed there are three countries we need to consider. In addition to the problems of Vietnam, which are well known, it seems that we now have to add those of Kampuchea (Cambodia) and Laos.

References

1. Westing, A. H. Herbicidal damage to Cambodia by J. B. Neilands, G. H. Orians, E. W. Pfeiffer, A. Vennema, and A. H. Westing, eds. in *Harvest of Death*, The Free Press, Macmillan Co., New York (1972), pp. 177–205.
2. Minarik, C. E., Shumate, J. E., Vakili, N. G., and Tschirley, F. H. A report on herbicide damage to rubber and fruit trees in Cambodia, Department of State document (12 July 1969).
3. George, A., Jr. Report of investigations, Re: defoliation damage in Cambodia, Department of the Air Force (21 August 1969).
4. Sihanouk, N., and Burchett, W. *My War with the CIA*, Penguin Books, London (1973), p. 258.
5. Shawcross, W. *Sideshow: Kissinger, Nixon and the Destruction of Cambodia*. Simon and Schuster, New York (1979).
6. Kissinger, H. *The White House Years*. Weidenfeld and Nicolson and Michael Joseph, London (1979).

7. Sambath, H. Letter to President of United Nations Security Council, 26 May 1969.
8. Sambath, H. Letter to President of United Nations Security Council, 17 June 1969.
9. Department of State telegram from U.S. embassy in Phnom Penh to the Secretary of State in Washington. September 1971. (declassified 17 July 1975).
10. Inouye, D. K. Personal letter, 19 December 1978.
11. Westing, A. Letter to W. Shawcross, 3 October 1979.
12. Westing, A. Personal letter, 17 June 1980.
13. Shawcross, W. Personal communication, October 1979.
14. LaBerge, W. B. Letter to Senator Max Baucus, 5 March 1980.
15. Baucus, M. Letter to Principal Deputy Under Secretary of Defense, 15 April 1980.
16. Hinkle, C. W. Letter to Senator Max Baucus, 23 October 1980.
17. Marchetti, J., and Marks L. *The CIA and the Cult of Intelligence*, Jonathan Cape, London (1974).
18. Church, F. Letter to E. W. Pfeiffer, 26 July 1971.
19. Turner, S. (Director, Central Intelligency Agency) letter to Senator Daniel K. Inouye, 23 June 1977.
20. Church, F. Personal letter, 29 December 1978.
21. Buckingham, William A. Jr. Operation Ranch Hand. The Air Force and Herbicides in Southeast Asia 1961–1971. Office of Air Force History, U.S. Air Force. Washington D.C. 1981 (unpublished).
22. Westing, A. H. Laotian postscript. *Nature (London)* **294**, 606 (1981).

9

Seveso

On Saturday 10 July 1976, an explosion at a chemical factory in the North Italian town of Seveso released a cloud of vapor which contaminated the surrounding area. The vapor, a chemical cocktail, consisted primarily of 2,4,5-trichlorophenol but also contained a quantity of the extremely toxic reaction by-product 2,3,7,8-tetrachlorodibenzo-p-dioxin[1] (dioxin).

The accident happened just after 12:30 P.M., $6\frac{1}{2}$ hours after the plant, owned by ICMESA, a subsidiary of the Swiss chemical company Givaudan and itself a subsidiary of the giant pharmaceutical corporation F. Hoffmann-La Roche, had closed down for the weekend. But the reactor and its contents had not been left in the usual condition. With the reaction run completed, company technicians had switched off the superheated steam used to heat the reactor and a few minutes later stopped the agitation of the vessel contents. The temperature gauge read 158°C. It was 6:00 A.M. and the end of a shift; the reactor operators reasoned that, left to its own devices, the reactor and its contents would slowly cool without requiring the addition of the several thousand liters of water normally used to stop the reaction and bring down the temperature of the mix.

It was a fateful decision. Six and a half hours later an unexpected rise in temperature resulted in a pressure increase sufficient to blow a safety valve on a vent pipe. The pipe projected above the roof of the building where the reactor was housed and was designed to protect the confines of the factory. In consequence, when the valve ruptured on 10 July, the reactor's toxic contents, which included some 250 g ($\frac{1}{2}$ lb) of dioxin, were discharged directly into the atmosphere and settled south of the factory over a residential area.

Local police were informed of the explosion by ICMESA officials on the day that it occurred. Residents living near the factory were warned by the officials not to eat any garden produce, as it was likely to be polluted. The police were asked to repeat the warning, but refused, claiming that they could only take instructions from the municipal health officer, and on that Saturday he was not to be found.

On the following day, Paolo Paoletti, ICMESA's director of production, was informed of the accident; he too could not be located on the previous day. Paoletti visited the Mayor of Seveso, Francesco Rocca, to explain the situation and to ask him—in the continued absence of the health officer, and his deputy—to instruct the police to issue notices warning against the consumption of fruit and vegetables from the area. Rocca agreed to pass on the advice and insisted that the mayor, Fabricio Malgrati, of the neighboring commune of Meda—in which ICMESA was sited—also be informed of the accident.

Two days after the accident, on 12 July, ICMESA's manager, Herwig von Zwehl, who had been informed of events of the previous day, conferred with Rocca and the local health officer, Francesco Uberti, to discuss the accident. Later in the day he confirmed their discussions in a letter to the health officer. The letter explained that the reactor was used to make trichlorophenol, implied that the chlorinated phenol was likely to be the major constituent in the vapor discharged from the vessel, but added that "we are not in a position to evaluate the substances present in the vapor or to predict their exact effects, but knowing the final product is used in manufacturing herbicides, we have advised householders in the vicinity not to eat garden produce."[2]

The letter made no reference to the possible presence of dioxin in the vapor discharge. Curiously, Von Zwehl also referred in his letter to the fact that trichlorophenol "is used in manufacturing herbicides." Givaudan never used the trichlorophenol for this purpose; all of the chlorinated phenol was sent to the company's plant at Clifton, New Jersey in the United States for conversion to the bacteriacide hexachlorophene.

On receipt of the letter, Uberti was satisfied that the fallout from the accident presented no risk to the residents of Seveso. He informed the mayors of Seveso and of Meda accordingly and sent a report of the accident to the provincial health office in Milan, 13 miles away. The letter arrived on 20 July, six days after it had been posted.[3]

Givaudan's senior management staff were informed of the accident the day after it occurred. Dr. Jorg Sambeth, the company's technical director and chief chemist, insisted that the reactor be sealed off and that samples be collected from the reactor and surrounding area for analysis. The analyses were to be performed by Givaudan's laboratories at Dubendorf in Switzerland. The laboratory routinely checked the trichlorophenol produced at ICMESA to ensure that the dioxin content was acceptably low and consistent with its ultimate use in the manufacture of a pharmaceutical product. In the event, analysis of the samples taken from Seveso would prove to be far more complicated and time consuming than the normal test. Substances which interfered in the assay of dioxin had to be removed from each sample before any analysis could be contemplated.

Dr. Bruno Vaterlaus, director of the Dubendorf laboratories, arranged for Sambeth's samples to be processed immediately. But the first results, Sambeth was told, would not be available before Thursday, 15 July. Meanwhile Vaterlaus agreed to accompany Sambeth to see the damage at Seveso for himself. Inspecting the damaged reactor with its charred contents, Sambeth began to be concerned that abnormally high amounts of dioxin might have been produced. These fears were realized the following morning when the first results from Dubendorf confirmed the presence of dioxin in quantities of parts per thousand, high by any standards. Sambeth immediately instructed ICMESA's manager to warn the Italian local authorities that toxic by-products might also have been vented in the reactor discharge. Givaudan's managing director, Guy Waldvogel, was also informed.

The situation at Seveso was clearly serious and Sambeth needed help to predict its outcome. Hoffmann-La Roche, the parent company of Givaudan, had to be informed of the accident but might also be able to provide the necessary information. At Roche's headquarters in Basle, Sambeth met and discussed the issue with Roche's clinical research director, Dr. Giuseppi Reggiani. Prior to the meeting, Reggiani had never heard of dioxin.[4] He agreed to help Sambeth and consulted colleagues for advice and information. The only literature on dioxin available discussed its toxicity in animals but not in humans.

To find out about the health risks, Reggiani phoned the plant doctor at Seveso, Dr. Ernesto Bergamaschini, for clinical details. Bergamaschini was reassuring about the health of the ICMESA workers; none had been affected. It was local children, he said, who appeared to have developed skin rashes. By the following Friday, more children were affected and Bergamaschini informed Reggiani, who in turn recommended that the children be hospitalized as a precautionary measure. Dr. Roberti, the local health officer for Seveso, had already arranged this. The two agreed to meet the following day in Milan and visit the children in the Mariano Comense hospital, near Seveso. Reggiani recalls seeing 12 children at the hospital, four of whom had very swollen faces.[4] (A report by the Lombardy Regional Health Authority[3]—later to assume responsibility for the health care of the Seveso residents—speaks of 19 infants being hospitalized.)

In view of what he had read about the toxicity of dioxin, Reggiani suggested that the four serious cases be moved to a hospital capable of providing intensive care in emergencies, in case the children's condition deteriorated rapidly. After much searching, a suitable place was found—the Niguarda Hospital in Milan agreed to admit them as urgent cases. Once at Niguarda, the four children were treated immediately with corticosteroids to reduce the water retention causing their facial swelling. By the following Sunday, there was a notable improvement with a reduction in the swelling. Reggiani claims that some of the doctors he saw were already talking about

"Agent Orange," a herbicide used in Vietnam and prepared from trichlorophenol which on occasions was heavily contaminated with dioxin (see Chapter 7). Could dioxin be present, they asked? "It is possible," Reggiani replied.

On Sunday 18 July Reggiani provided the doctors at Niguarda with copies of the literature he had collected dealing with dioxin. He recommended that in addition to the regular tests carried out on the children, they should undergo liver and kidney function tests, a blood count and a check for the presence of porphyrins (pigments) in the urine, as these were parameters known to be altered by dioxin in animal tests.

Meanwhile at Seveso the general situation had been deteriorating. On Wednesday 14 July there were the first deaths from among the animals poisoned by dioxin. Reports of the deaths reached Mayor Rocca's office. The Mayor became increasingly alarmed, reflecting the concern felt by many Seveso residents. Acting in concert, the Mayors of Seveso and Meda issued orders on 15 July declaring an area south of the ICMESA factory to be polluted and banning the consumption of local fruits and vegetables. On Saturday 17 July the ICMESA factory was pronounced polluted and orders were issued for the destruction of all vegetable products, fruit, and animals in the contaminated area. Closure of ICMESA occurred on the following day, 18 July, by order of Mayor Malgrati of Meda. To prevent any tampering with the damaged reactor, seals were placed on the access doors to the plant on the instructions of a local magistrate; the reactor was now under the jurisdiction of the court.

Although Mayor Malgrati's order had closed ICMESA, the management had already brought the plant to a standstill on Friday, 16 July. Workers at the plant, incensed that they had not been taken into the company's confidence and given details about the accident, decided to strike to force the management's hand. Two months earlier they were forced to resort to similar action before the management would agree to let the regional factory health inspectorate enter the plant to assess health hazards. The workers had asked for this assessment for two years, but without success. The company continued to prevaricate until the strike in May 1976 when it had no alternative but to accede to the workers' demands. By striking again on 16 July the workers were hopeful that ICMESA's manager Von Zwehl would be more forthcoming. But Von Zwehl had nothing to say. He claimed he knew no more than he had stated five days previously in his letter to the local authorities describing the accident. Von Zwehl's apparent ignorance of events of ICMESA was not well received by Salvatore Adamo, the magistrate from the neighboring town of Desio who had sealed off the trichlorophenol reactor. On 21 July, Adamo had Von Zwehl and his technical director Paolo Paoletti arrested on the grounds of causing a "culpable disaster."

On the same Saturday (17 July) that Reggiani was visiting the Seveso children the story of the accident appeared on the front page of the Milan newspaper *Il Giorno*. The story had broken. In a matter of hours, Seveso was inundated with reporters. Events now took place in an intense glare of publicity. The issue was a serious matter and for that reason the limelight remained fixed on Seveso for months on end. Seveso would eventually become a watchword for pollution.

Reggiani returned to Roche's headquarters in Basle on Tuesday, 20 July to begin a series of phone calls to other chemical companies known to have had accidents involving trichlorophenol. Reggiani says he contacted Dow Chemical in the United States, Coalite in Britain, Philips-Duphar in Holland, BASF in Germany, and Chemie Linz in Austria. The answer they all gave was identical: the population of Seveso had to be evacuated as soon as possible. But for Reggiani there were some practical problems to be considered. How many people were at risk and should be evacuated?

The answer to this lay in the analyses of the contaminated ground and vegetation then being carried out in Givaudan's laboratories at Dubendorf. Dr. Bruno Vaterlaus, in charge of the investigation, was attempting to construct a map of the contaminated area. By the evening of 20 July Vaterlaus had the map with 36 locations marked as dioxin contaminated. The map, together with some pure samples of dioxin, was handed over to Dr. Aldo Cavallaro, Director of the Provincial Laboratory of Hygiene and Prophylaxis (PLHP) in Milan, and Dr. Guiseppi Ghetti, the municipal health officer for Seveso and Meda.[5] The two men had been sent to Dubendorf for the details of the laboratory analyses. Over the weekend of 17/18 July the Lombardy Regional Government, responsible for the administration of the Seveso area, had begun to take stock of events in the contaminated region to the north of Milan. Regional officials were responsible for the health of the Seveso residents and an official report of the Regional Government's Health Assessor's Council (HAC) says that on 18 July Givaudan "does not admit that there is dioxin (at Seveso); it allows this to be understood." Cavallaro, who claims to have suspected the presence of dioxin in the discharge from the reactor on 18 July (the same day that Reggiani was discussing the possibility with Italian doctors), and Ghetti were thus sent to Dubendorf with the blessing of Dr. Vittorio Carreri, the provincial health officer. Carreri, and his superior, Dr. Vittorio Rivolta, Minister of Health for Lombardy, had to know whether dioxin was indeed a threat to the Seveso residents. The HAC in its report says that Cavallaro and Ghetti received confirmation from Givaudan of the "qualitative presence of TCDD [i.e., dioxin] outside the (ICMESA) factory."[3] Givaudan, on the other hand, claims that the information it handed to the two Italian visitors outlining the extent of the contamination was sufficient to indicate that there was a problem of considerable magnitude at Seveso.

Cavallaro and Ghetti returned to Milan to report back to the regional officials on their discussions with Givaudan. With his pure dioxin samples Cavallaro began a frantic program of investigation to confirm Givaudan's findings that dioxin was indeed present in the soil and vegetation at Seveso.

As the days passed the situation at Seveso deteriorated further. On Tuesday 20 July another eight people with skin inflammation and complaining of vomiting were hospitalized at Niguarda after handling, and presumably eating, contaminated fruit and vegetables. Carreri in his office in Milan was anxious to keep in touch with developments and to warn local doctors to look out for certain types of illnesses; he drafted a circular to them stating that they must report all cases of gastrointestinal disturbances or skin rashes. The circular was sent out as a telegram.

However, circulars on their own were clearly not enough to contain the situation. Following a meeting with local health authorities at Seveso on Wednesday 21 July the regional Minister of Health, Dr. Vittorio Rivolta, insisted that some urgent measures be taken to prevent the spread of contamination. Local authorities in the neighboring communes of Cesano Moderno and Desio announced restrictions on the consumption of fruit and vegetables. All traffic on two of the major roads in Seveso was banned. A commission of experts to assess the health implications for the Seveso residents was set up. By Thursday 22 July Rivolta's instructions had been enforced and a more detailed analysis of the dioxin contamination began. To minimize the risk to children Mayor Rocca requested and was granted permission by the regional authorities to send 90 of the youngest to a convalescent camp. The outbreak of skin rashes was an early visible indication that the local population had been contaminated by the reactor discharge. A watching brief on the skin complaints was an urgent priority and arrangements were made to open a dermatology clinic under the supervision of Professor Vittorio Puccinelli of Milan University.

Friday morning, 23 July witnessed the first large-scale meeting of officials to discuss the Seveso problem. The occasion was an extraordinary meeting of the Provincial Health Council to which leading scientists and doctors were invited. Academics from Milan University, scientific institutes in Milan and Rome, and the Italian Ministry of Health were all in attendance together with local health officers and most of the mayors representing the communes which had been affected by the pollution. Cesare Golfari, the president of the Lombardy Regional Government, presided over the meeting.

For Rivolta, the extraordinary meeting was an opportunity to explain developments at Seveso and the action taken by himself and his officials. He hoped to convince his colleagues that the situation had been dealt with correctly and efficiently. The meeting would also give some authority to his actions providing he could win the approval of his audience for his proposals. Rivolta was successful. The assembled experts concurred with his decisions. The measures adopted at Seveso to protect the public from dioxin were deemed adequate. There was no need to evacuate the population; everything was considered to be under control. In a television interview later that day, Rivolta repeated this message. He said viewers could rest assured that the situation at Seveso was under control.[6]

Lombardy's Minister of Health was to regret giving that interview. Beforehand, an attempt had been made to inform Rivolta of the seriousness of the situation at Seveso. Guiseppi Reggiani, Roche's clinical research director, had tried to intervene at the extraordinary meeting and make a statement. His attempt was rebuffed. He was informed by the vice prefect that the meeting was not open to the public, and he was asked to leave. Before doing so, Reggiani left a message for Seveso's health officer, Guiseppi Ghetti, stating that he had important information and must talk both to him and to Rivolta. The information that Reggiani had to convey was that the population at Seveso had to be evacuated without delay; this was the opinion of everyone he had spoken to with experience of accidents involving dioxin.

Reggiani and Rivolta eventually met at Seveso town hall on the evening of Friday, 23 July. Rivolta had been presiding at a meeting of local administrators held to finalize the details of the public health measures and to define the extent of the contamination on the basis of the location in which animals had died and vegetation wilted. Rivolta was satisfied with what he had achieved. His satisfaction was not to last; when he met Reggiani, the latter's blunt dismissal of his public health measures and his insistence on the need to evacuate the population shook Rivolta. That morning Rivolta's own experts had concurred with his opinion that the Seveso residents need not leave their homes.

In the urgency of the moment, Reggiani had traveled to Italy without any written credentials. Rivolta challenged him on this. On whose authority did Reggiani speak? Where was his proof that the population should be moved? In reply Reggiani claimed that he spoke for himself, adding that he had the authority to commit the company to this course of action. Evacuation, he said, was necessary on medical grounds. An angry Rivolta said that he could reconvene the meeting of experts for the following morning—Saturday, 24 July. Reggiani, he insisted, would present himself suitably credited and with the information necessary to support his demand.[7]

Reggiani, however, did not attend that meeting. Bruno Vaterlaus, Director of Givaudan's Dubendorf laboratories attended in his place handing over a letter from the company's director, Guy Waldvogel, together with a more comprehensive map of the extent of dioxin contamination at Seveso than had been handed over to the representatives of the Lombardy Region who had visited Givaudan's laboratories four days earlier. The new map gave details of dioxin contamination of soil and vegetation at 44 locations.

Vaterlaus had agreed to stand in for Reggiani at short notice. In his determination to reconvene the panel of experts, Rivolta asked the police to locate everyone who had attended the Friday meeting. Reggiani was to have been guarded too to ensure his attendance. Tipped off by a colleague that the police were looking for him Reggiani decided to leave Milan and return to Switzerland. Givaudan's managing director Guy Waldvogel had been apprised of Reggiani's difficult position and agreed to meet him in the Swiss lakeside resort of Lugarno. Bruno Vaterlaus and Dr. Jorg Sambeth, as the two Givaudan scientists most involved with the accident, were also asked to convene at Lugarno. Between them the four composed the letter to be delivered to Dr. Rivolta insisting that the people at Seveso should be evacuated.

Rivolta had taken steps to ensure Reggiani's appearance at the meeting to be held on Saturday, 24 July. But as far as Reggiani was concerned his job was done. He had no need to return to Milan. Vaterlaus could go in his stead.

There was other evidence besides Vaterlaus' map for the Saturday meeting to consider. At 11:00 P.M. on Friday evening Dr. Cavallaro phoned Rivolta and his deputy Carreri to say that Givaudan was not the only group which had identified dioxin in samples from Seveso; he too had also identified the chemical. With the help of colleagues at Milan University's pharmacology department, Cavallaro had analyzed material collected at Seveso. In a report prepared by the Lombardy Region it is stated that Cavallaro's information was the "first quantitative data" on the dioxin concentration at Seveso. Givaudan and Roche dispute this. They claim that Cavallaro had already been given the results for 36 analyses when he visited the Dubendorf laboratories. When he phoned his results through Dr. Cavallaro probably had information on only 2–4 samples. When asked how many measurements he had made he declined to say.[5]

The evidence from Vaterlaus and Cavallaro was quite clear, however, and suggested one course of action—evacuation. The population at Seveso had to be moved without delay. Waldvogel in his letter to Dr. Rivolta insisted that this was the only proper course.[8] Rivolta's colleagues, reassembled to hear this verdict, reversed the decision of the previous day and ordered the evacuation. Two days later on 26 July the first 225 residents left their homes (170 from Seveso, 55 from Meda). On 2 August a further 511 people were evacuated.

All the residents ordered to leave their homes were from a region of Seveso and Meda considered to be the most heavily contaminated. This 269-acre, triangular-shaped area extending south/southeast of the ICMESA factory was designated Zone A. Surrounding it was a 669-acre, less contaminated site, in which the residents were allowed to remain. Zone B, as it was called, extended the area in "dogleg" fashion further to the south (see Figures 16 and 17). Finally, an area bordering on Zone B was classified as an observation region and often referred to as the "zone of respect," Zone R. Dioxin contamination in the 3575-acre Zone R was low level and patchy.

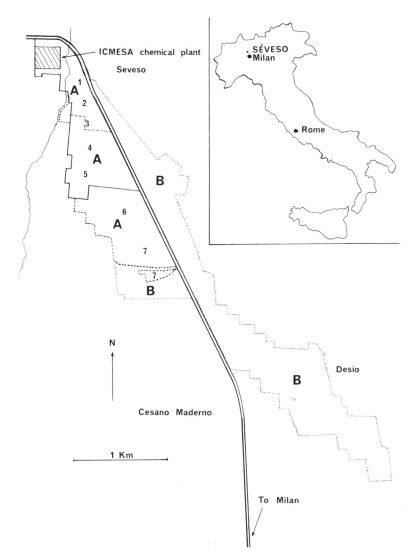

FIG. 16. Zones A and B at Seveso. Source: Hay, A. *Nature* (*London*) **267,** 384 (1977).

In ordering the evacuation of the Zone A residents, but not that of those residents in Zone B, the health authorities were taking a calculated risk. Housing the 736 residents from the most contaminated region was difficult enough, both for the authorities and the residents. Inevitably the temporary accommodation in hotels near Seveso and Milan was cramped and some families had to be split up. To make matters worse, the evacuated Seveso residents were treated by others as though they themselves were contaminated with dioxin and spreading the chemical wherever they went.

The difficult circumstances for the Zone A residents were much to the fore in the minds of the health authorities in Milan as they considered what to do about the residents in Zone B. Moving them would have entailed housing a further 4800 people. There were political risks involved whatever decision the authorities decided to take. Moving people would increase their discomfort, leaving them would expose them to dioxin. Roche and Givaudan had insisted that the Zone A residents be evacuated even though the amounts of dioxin to which they were exposed were probably far lower than workers had been in contact with in industry. The companies had acted with prudence. For Rivolta and his colleagues in the Lombardy health department the prudent course would be to allow the Zone B residents to remain in their homes. They prayed that the companies would not insist on evacuating Zone B. Fortunately, for Rivolta, Roche and Givaudan did not ask for this.

But this was not the end of the matter for the Lombardy Health Authorities. The people in Zone B were still at risk from the "lower" dioxin concentrations. Two groups in particular were most at risk—children and pregnant women. Most of the burns which appeared in the first fortnight following the accident were observed in children, and suggested that they ought to be treated as a special category. Contingency plans were therefore drawn up to facilitate their removal from the area for the daylight hours.[9]

Dioxin's proven ability to cause malformations in animals implied that it might also be a teratogen in humans. The developing fetus is most at risk from teratogens during the period of organogenesis when tissue and organ differentiation occur. In the human fetus this period generally occurs between gestation days 18–55. It was considered advisable, therefore, that women in the first trimester of pregnancy should move from the area until the fetus was at least three months of age.

For the remaining residents appeals to abstain from procreation were issued, as were orders from the Lombardy authorities forbidding the consumption of fruits and vegetables from the area. Crop planting was forbidden and residents were advised to wash regularly. To ensure that no contaminated animal products were consumed it was decided that all domestic animals in the area should be killed.

Inevitably, some of these rules were flouted. Increasing school hours for the children kept them off the contaminated Seveso soil for most of the day, but it was impossible for parents to keep their children housebound at all times. Chickens and rabbits were hidden from prying eyes by some residents reluctant to begin slaughtering before the animals were ready for the pot. In general, however, the inconvenience caused by these measures did lessen peoples' contact with the polluted soil and vegetation.

During evacuation everyone was instructed to take only the minimum of clothing, preferably items which had been stored for a few weeks in drawers and wardrobes; everything else was to be left behind. People clearly had differing ideas about a "minimum" quantity and many therefore took as much as they could carry, for they had no idea how long they would be away. To protect the homes of the occupants the authorities initially forbade all entry into Zone A by outsiders. But then, according to Thomas Whiteside in his book *The Toxic Cloud and the Pendulum*, the authorities "changed the rules to allow ten persons to enter on weekdays and twenty on Sundays."[10] As Whiteside points out the illogicality of this was obvious to all. When people in the area heard about the safety measures they just laughed. Their arithmetic was simple, for Whiteside notes that "they would say to one another 'so dioxin is twice as dangerous on weekdays as it is on Sundays?.'"

Much harsher criticism of the local authorities responsible for labor, air pollution, factory inspection, and the fire brigade was made by the Parliamentary Commission of Enquiry, set up on 16 June 1977 to investigate the Seveso accident. Commenting on the events leading up to the accident, the Commission reported that there was insufficient communication between the various authorities even though provision had been made for dialogue.

The Commission said that the authorities took the delineations of their respective functions too literally and, in consequence, were too insular, giving too little thought to issues of safety. It was the Commission's view that had regular contact between the authorities occurred, and had this been coupled with more inspection of the ICMESA trichlorophenol reactor, the accident might have been averted because additional safety features would have had to have been installed on the reactor.[11]

The Commission also had some harsh words to say about Givaudan's record of communication with its workforce and on the inadequacy of the design of its trichlorophenol reactor. The company admits that the workers at its ICMESA reactor were not fully informed about the hazards of dioxin.[5] But Givaudan disputes that its reactor design was at fault and insists that the reactor had been constructed after taking due account of previous accidents involving trichlorophenol reactors in other countries.[12]

Rather surprisingly, however, the Commission—in an otherwise unbiased report—gives an inaccurate account of the two-week aftermath following the accident on 10 July. The chronology of events between 10–24 July 1976 given earlier in the chapter is based on internal documents prepared by the Lombardy Regional Health Board and on discussions with individuals involved with events at Seveso at the time. It is clear from the Lombardy Region's own documents that Givaudan and Roche provided much of the detailed information and results which formed the basis for the ultimate evacuation of the Seveso population. The Parliamentary Commission omits to mention this fact, or to say that Italian scientists were in possession of a good deal of this information at the same time that the Lombardy Health Authorities were assuring the residents of Seveso that they faced no health risk. A press version of the Parliamentary Commission's report issued months before the full report was available was even more misleading in its implication that it was results from Italian laboratories in the detection of dioxin, and not those from Givaudan, which were used as the basis for ordering the evacuation.[13] This was quite untrue. It was hardly surprising that Givaudan and its senior partner, Roche, would take umbrage over this section of the Parliamentary Commission's report, which one Roche senior executive described as a "twisted tale."[5] Roche knew that the report was not accurate in describing the company's role at Seveso and it was anxious to make this widely known. It was important for the company's image that it should be seen and acknowledged to be doing as much as it could at Seveso. The company had had too many public relations failures in recent years to let this latest issue pass without comment.

A multinational company with a turnover of some £2000 million, Roche began life officially in 1896 when one Fritz Hoffmann became owner of a small chemical company in Basle. Hoffmann recognized the potential gains to be had in producing pharmaceutical products of proven reliability. Together with a colleague Emil Barell, a chemist, Hoffmann began a rapid expansion of the company, acquiring new companies abroad and diversifying the range of its products. Hoffmann-La Roche became well known for the production of vitamins and its predominance in this field was established in the 1930s and 1940s as company scientists succeeded in synthesizing one vitamin after another and adapting the synthesis to large-scale production. As for the vitamins Roche itself had not identified, the company was quick to establish relationships with the scientists who had made the initial breakthrough.[14] In this way Roche rapidly became the predominant world producer of vitamins, a status it guards with some degree of ruthlessness.[6]

Prior to the Seveso accident, Givaudan, Roche's junior partner, had few complaints about the publicity it attracted. Begun as a family concern in Geneva in 1894 by two brothers, Xavier and Leon Givaudan, the company rapidly established itself as a producer of perfumes and flavors. Today the company's reputation is still based on its preeminence in these fields.

in this context. The absence of the skin disease does not preclude exposure to dioxin, but suggests that contamination by the chemical was mild.

Sight impairment is said to be a consequence of exposure to the 2,4,5-T-based herbicide Agent Orange. Reports from Vietnam indicated that a degree of eye damage was observed in subjects exposed to the herbicide during U.S. military defoliation programs during the Vietnam War. The same problems were not observed at Seveso. Of the 400 Seveso residents who were examined by oculists, none was said to have any pathological symptoms attributable to dioxin.

Any abnormalities in liver function can be detected by measuring relevant enzymes in serum. Tests performed on individuals from Zones A and B indicated that over a period of a year, 10% had raised serum enzyme levels, a finding consistent with some degree of abnormal liver function.[20] The enzyme changes, elevations in serum transaminases and gamma glutamyl transpeptidase, were more common in patients from Zone A than those from Zone B. It is difficult to impute some of these changes to dioxin without other measurements. In a review of the clinical assessment of the Seveso residents Professor Nicola Dioguardi of the Milan University's Institute of Clinical Medicine insists that information on other serum parameters—the enzyme alkaline phosphatase and the pigment bilirubin—are necessary before liver toxicity can be assessed. Biopsy samples of liver for examination under the microscope are also necessary in Dioguardi's view to confirm the diagnosis of liver damage, and further investigations are needed to clarify the situation, she says.[20]

One of the better known effects of dioxin poisoning is a reduction in the effectiveness of the immune system. Seveso residents were, therefore, monitored for any changes which might be consistent with dioxin exposure. Professor Gaetano Fara, chairman of the Health and Epidemiology Commission stated in 1977 that a report on the children from Seveso who had been seen by doctors on three separate occasions had revealed no immunological damage in this group compared with a matching peer group in a noncontaminated area.[21] Addressing a conference in Rome in October 1980, Fara reported that one test used to assess the immune system—measurement of lymphocytes (white cells) in blood—had indicated higher values than normal among Seveso children. Serious dioxin exposure should have caused a fall in lymphocyte counts.[22] The reason for the changes was unclear, said Fara. He added, however, that the population would continue to be monitored for immunological changes for some time yet and that perhaps with time the findings would be clearer.

Every cell in the body carries its genetic code on chromosomes and one effect of dioxin poisoning in animals is chromosome damage. The implication of this disruption is not clear. Some scientists believe that there is a link between chromosome damage in somatic cells—those not involved in reproduction—and cancer, but this has yet to be proved. Tests for chromosome damage have been performed on Seveso residents; 34 women from the contaminated zones at Seveso were able to obtain legal abortions in Italy following their exposure to dioxin. Twenty-three of the fetuses were examined for chromosomal abnormalities by Dr. Luigi de Carli at the University of Pavia. He says the incidence of abnormalities was no greater than the accepted norm.[23] Similar negative findings were obtained in a study performed on 301 residents from Seveso. According to de Carli there were no significant differences between these individuals and the 87 people who constituted the control group. But de Carli is still not convinced that the inhabitants of Seveso are in the clear and he recommends that they still be monitored[23] for any immunological disturbances which may appear in the future.

Some of the changed blood parameters of the Seveso residents are difficult to attribute unequivocally to dioxin exposure. No such difficulty is encountered concerning the outbreak of skin disease in the area following the reactor discharge: 187 children developed the disfiguring skin condition, chloracne, as a direct result of their exposure to dioxin. Some children were directly exposed to the vapor cloud as it condensed and settled over Seveso. Only days after their exposure, the children developed extensive skin rashes and a marked puffiness of the face. The rapid onset of their condition precluded the effect of dioxin—which causes blackheads and cysts weeks after contact with the chemical—and by common consent the children's skin condition was attributed to burning by 2,4,5-trichlorophenol, the principal constituent of the vapor cloud (see Figure 18).

Treatment of the burns and swelling was effective; both conditions improved and by mid-August 1976 had healed. Some 5–6 weeks later, at the end of September, the much more persistent condition chloracne developed.[24] Blackheads began to appear on the children's faces, particularly around the eyes (see Figure 19). From September to December 1976, 50 children developed the condition. In subsequent screenings of over 32,000 children in nursery, primary, and secondary schools in 10 communes including Seveso between February and April 1977, a further 137 children were found to have the skin complaint.[22] The program to identify chloracne was organized by Professor Vittorio Puccinelli, the director of Milan's Dermatology Clinic. Commenting on the reasons why only children were affected by the condition, Puccinelli said that three factors were concerned: the structure of a child's skin differed from that of an adult's; children were outside more often and exposed as such to more dioxin; and they were usually less careful about hygiene.[24]

By early 1978 it was apparent that there was a marked improvement in the severity of the chloracne in all the children, and only six new cases

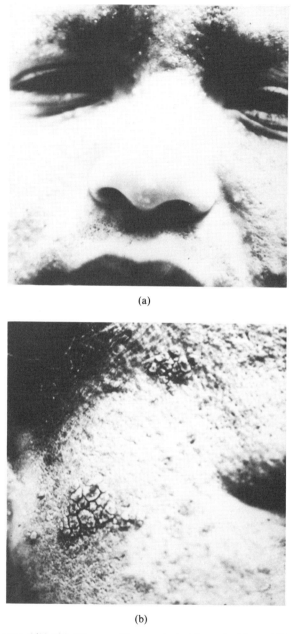

FIG. 19. a. Seveso child with chloracne. Face marked with small white and black comedones (similar to blackheads). b. Seveso child with chloracne. The face is marked with white and black comedones (raised spots) and there are areas of the skin with a thick, hard, horny layer of keratosis.

had been identified.[25] Two years later, no new cases of chloracne had occurred and according to Senator Luigi Noe, the head of the Lombardy Region's Special Office for Seveso, those children who had the skin condition no longer have it. "It has cleared," says Noe.[26]

For some residents of Seveso there were psychological scars which will take far longer to heal. This is especially true for women who were pregnant at the time of the dioxin discharge, or who became pregnant in the ensuing weeks. Dioxin is known to cause malformation in the offspring of experimental animals; there was clearly a risk that it might also affect the developing human fetus. With this in mind the Lombardy Regional Health Council adopted a resolution at its meeting on 2 August stating that women residents of the dioxin-contaminated area should be provided with detailed information on possible risks and offered medical assistance to enable them to assess the situation. The Council also stated that existing laws made provision for the women to have abortions should they so decide.[27]

Abortion was made legal in Italy following a Constitutional Court ruling in February 1975, permitting the operation in circumstances where the mother's physical and psychological health was considered to be at risk. The court ruling replaced a penal code, established by Mussolini in 1932, which ruled that abortion was a crime against the health and integrity of the Italian race. However, in spite of the Lombardy Regional Authority's instructions to local physicians in Seveso and neighboring communes that the court ruling should be observed, women in the area encountered considerable hostility from doctors, local politicians, and the Catholic church, all of whom had pronounced conservative views on abortion. Cardinal Colombo, Archbishop of Milan, preached about the evils of abortion and appealed to women to see their pregnancies through to term, even if it meant the birth of a malformed infant. There were families who would adopt the children if their own parents did not want them, Colombo said, and he praised couples who had volunteered to adopt malformed children born in the Seveso area. The Archbishop appealed to others to make similar offers. But it was the Vatican newspaper *Osservatore Romano* which removed all doubts with its rhetorical question "Is the mere possibility of the risk of malformation a sufficient reason to eliminate an unborn human being which undoubtedly has a greater chance of being born normal than abnormal?"[23]

Many local obstetricians and gynecologists did not need to ask themselves this question. For them the answer was no,[28] and they attempted to persuade the Seveso women accordingly. Persuasion took many forms and it was not unusual for women to receive distressing treatment, including being forced to listen to the heartbeats of the fetus with a stethoscope.[29] In some cases the actions of the local doctors was sufficient to dissuade women from seeking to terminate their pregnancy. Most Seveso women did not even consider having an abortion. For those who decided that the risk of giving birth to a malformed child was too great, there were several options. They could try and insist on having a legal abortion in Italy or alternatively procure an illegal abortion in Italy or abroad. According to the Radical Party Member of Parliament for the constituency of Genoa, Ms. Adele Faccio, many women had to resort to illegal abortions. Having been in the Seveso area in July and August of 1976 Faccio was concerned about the question of abortions for the Seveso women. According to Ms. Faccio, only 30 women were permitted legal abortions in Italy, "whereas twice this number have had illegal abortions. The illegal terminations were performed by "women's self-help groups" for cases where the fetus was not older than 10 weeks. For women in a more advanced state of pregnancy it was essential to send them abroad for an abortion."[21] Women from Seveso are known to have traveled to Switzerland and Britain for the operations.

Most of the legal pregnancy terminations in Italy were performed at Milan's Mangiagalli Hospital for Women under the guidance of Dr. Francesco d'Ambrosio. The Lombardy Region's resolution on 2 August calling for medical help to be given to the Seveso women was fine in theory. The question was who would provide this help? Ambrosio took the decision to open the clinic and to offer the women advice on birth control. Hostility to the decision was swift in coming from local right-wing politicians and the clergy; but two years later the actions of d'Ambrosio would seem less controversial. Following the passing by the Italian Parliament of a more liberal abortion law in 1978, abortion is now legal in Italy and free of charge for women less than 90 days pregnant.

But if legal abortions proved difficult for the Seveso women there was some relief in Italian medical circles over the results of the examinations performed on the women permitted legal terminations. The fetuses examined appeared to be normal. But the report of the examining panel of doctors, Professor Alfred Gropp and Dr. Helga Rehder of Lübeck Medical School, West Germany, and Professor Francesco Cefis and Dr. Laura Sanchioni of the Istituti Clinici di Perfezionamento in Milan, was qualified. The panel pointed out that of the 34 fetuses—30 from women who had legal abortions, four from women who had experienced miscarriages—examined, one showed signs of some abnormal development. As with most of the fetuses examined, however, this one embryo was badly macerated—a common situation with induced abortions.

The results of the study according to Gropp and his colleagues had to be interpreted with great caution. There could be a suggestion in this one instance of the existence of Down's syndrome. This, however, is not an uncommon abnormality in most populations. In a report to the Lombardy Regional Health Authority Professor Gropp and his colleagues point out that "the results of the embryo-morphological studies do not have any hint for damage brought about by the action of exogenous agents (dioxin)." They added the rider, that the negative result did not necessarily indicate that dioxin "does not bear a risk for the mother and fetus."[30] The examinations carried out by the four scientists had certain inherent limitations; minor malformations and particularly organ damage are not detectable in the very young fetus, only a limited number of cases could be studied, and most of these embryos were incomplete and of different gestational age.[31]

Professor Gaetano Fara, chairman of the Health and Epidemiology Commission for Seveso, added a further qualification to the interpretation of the results. He admitted that the examinations of fetal material could not be interpreted in the context of a true epidemiology survey; it represented an analysis of the limited data available. Fara added that, for a true epidemiological survey, the population would have had to have been much larger and the study more carefully controlled.[21]

Interpreting the evidence on birth malformation rates at Seveso is even more difficult. By June 1977 eight women residents in the dioxin-contaminated area had given birth to malformed children. Commenting on the figures, Professor Fara claimed that statistics would indicate that this incidence of malformation is no greater than the average for the region,[21] following figures which Fara had himself obtained from a survey of birth defects in Italian cities.[32] The official figures for birth defects in the Seveso region (11 communes including Seveso) are four congenital malformations out of 3902 births in 1976; 38 malformations from 2774 births in 1977, and 53 malformations out of 2777 births for 1978.[33]

The striking feature of the malformation rates is the increase from 1976 to 1978. This is uniformly regarded as being due to better reportage.[34,35] Having been made aware of the risks posed by dioxin, physicians became more reliable in reporting any birth abnormalities which they had observed. Prior to the reactor accident there had been gross under-reportage of birth deformities. Without a national register for recording the cases there was no incentive for doctors to report them.

In spite of the apparent increase in the reported incidence of abnormalities at Seveso the rate of malformation for the region is no higher than that for Western Europe. The average incidence of birth deformities for Western Europe is 2.5–3.0 for every 100 births, i.e., 2.5–3.0%. Fara[32] has reported a figure of 2.97% for Milan and 2.32% for Italy as a whole. The figures for Seveso are thus 0.13% for 1976, 1.36% for 1977, and 1.9% for 1978 and, as such, well within the limits for birth anomalies in Europe. Furthermore it has been noted that there is a wide variation in the type of abnormalities observed and this is also considered to be normal.[34] Most agents which have been proved to cause fetal abnormalities produce one or two characteristic defects, such as the limb deformities observed in children whose mothers took the sedative Thalidomide. Such a pattern has not been observed in the Seveso children. There is no striking increase in any one type of abnormality which could be attributed to exposure to dioxin.[34]

But the issue is by no means resolved. Arguments persist at Seveso over the type of abnormalities which should be recorded. There is also the question of whether abnormalities which may be detected later in a child's life should be included in the register of birth defects. If the register were to include minor birth defects—those which are of no serious medical or cosmetic consequence to the patient—the number of anomalies increases considerably. For example there were some 89 additional anomalies in 1978 not officially notified.[33] It is not yet clear, however, whether the overall picture would alter if this procedure were adopted. According to Professor Marcus Klingberg, chairman of the International Committee of Experts—an advisory panel of international scientists of world renown appointed by the Lombardy Regional government—it is still not possible to reach "any definitive conclusions as to whether or not any of the congenital malformations exceed the expected rate. Malformations that at first glance appeared to exhibit very high rates, namely microcephaly, hypospadias, hemangiomas and Down's syndrome, must be further analyzed to exclude the possibility that they may have occurred by chance . . . or that they are within the normal baseline rates for Italy."[36] Monitoring minor birth defects is important, Klingberg says, because they might "reflect the effects of environmental insults to the developing fetus."[36]

Views about the significance of the rate of miscarriages, or spontaneous abortions, at Seveso are equally divided. According to the World Health Organization fetuses which are spontaneously aborted often show some degree of malformation. Genetic defects, infection, trauma, and certain drugs are a few of the agents known to cause spontaneous abortions.[37]

According to the Lombardy Region's Special Office for Seveso the overall incidence of spontaneous abortions for Seveso and ten surrounding communes changed little between 1973 and 1977. Some 10.89% of all pregnancies were spontaneously aborted in 1973; the comparable figures for 1976 and 1977 were 9.74% and 11.63%, respectively.[38] The increase in 1977 is thought to be due to more accurate reporting of incidents.[38] If the incidence of spontaneous abortions is assessed in a different way, however, the picture is considerably altered. The rate of spontaneous abortions is often calculated over a calendar year by relating the number of pregnancies resulting in miscarriages in the year to the total number of live births for the period. An alternative procedure would be to relate pregnancies which aborted spontaneously to those which go to term and which were conceived at the same time. Epidemiologists at the Seveso Special Office argue that this is a more reliable method which could alter the incidence of spontaneous abortions. Using this alternative method the Seveso Special Office says that the rate of spontaneous abortions rises in the fourth trimester of 1976 in towns where over 50% of the inhabitants live in the dioxin-polluted zone.[38] The office claims that with reference to two towns in the contaminated zone, Cesano Moderno and Seveso, the rates for the trimesters July–September 1976, October–December 1976, and January–March 1977 were 16.5%, 21.3%, and 11.2%, respectively. The well-known French teratologist Dr. Tuchmann-Duplessis doubts that this rise is causally related to dioxin. He has two criticisms of the results. Firstly he says the incidence of spontaneous abortions reported for the area is said to be well below the norm for Western Europe and hence cannot be a true reflection of developments. Secondly, he claims that there is no evidence to show that the rate of spontaneous abortions is higher in the most contaminated zone than in those areas least polluted.[34] Had dioxin caused an increase in spontaneous abortions, the rate should have been highest in the most contaminated zone. This is unlikely to be the last word on the subject as further consideration is given to the evidence available.

Dioxin has been shown to cause cancer in animals and the Seveso Special Office has, therefore, decided to establish a cancer registry for residents in the area. If the chemical has any affect on the rate of cancer this is unlikely to be noticed for many years to come in view of the long latency period between exposure to chemical carcinogens and the development of tumors.

The effect of dioxin release at Seveso is still much in evidence in other ways. Most of the land heavily contaminated with the chemical and designated Zone A is still uninhabited. Surrounded by a barrier of yellow plastic sheeting and enclosed by a seven-foot-high wire fence, Zone A is only entered by employees of the Lombardy Region engaged on the decontamination program for the area and by scientists investigating the persistence of the chemical in the soil.

It is this very persistence which is causing difficulties for the authorities. Initial measurements in the area suggested that the concentration of dioxin in the soil would fall by half over a period of 2–3 years.[39] This is now known to be untrue. According to Dr. Vito La Porte, a research chemist employed by the Lombardy Region, there is little difference between recent dioxin measurements in Zone A and those made several months, or indeed two years ago.[40] The dioxin located in the top soil at Seveso is not being degraded, says La Porte. Other scientists at the Istituto Superiore di Sanita in Rome note that, if the chemical *is* breaking down, then the rate of degradation is very slow. Dr. Allessandro di Domenico and his colleagues estimate that it will take more than ten years for the present concentration of dioxin in the top soil to fall by half.[41]

Estimates of the amount of dioxin discharged form the ICMESA reactor vary but are generally considered to be in the range of 240–500 grams.[21,42] This initial concentration is said by di Domenico to have fallen by half in the five-month period after the accident, when the chemical was on the soil surface and exposed to sunlight—one agent known to degrade dioxin. But after a while the chemical although almost immobile in soil did percolate into the first few inches of topsoil where it has remained ever since. Once below the soil surface the rate of degradation has slowed considerably. The Italian authorities are faced with devising a solution for the containment of a chemical which could be present for many, perhaps hundreds of years.

According to Senator Luigi Noe, in charge of the Lombardy Region's Seveso Program, one of the remaining options is to dig a large hole and bury the contaminated top soil.[26] Plans existed to excavate an area 100 × 100 meters to a depth of 2–3 meters. An estimated 50,000 cubic meters of top soil is to be deposited in this basin which is to be lined with clay and plastic, capped with the same materials together with a layer of sand and reinforced concrete. The containment will be permanent says Noé. "It is" he points out "the same philosophy as that for burying nuclear wastes."

Burial will not be practical for all the contaminated soil says Noé. "It is just too expensive." For this reason, scientists are experimenting with different procedures to encourage the breakdown of the chemical. They have tried inverting the soil to expose dioxin to the elements, but this has been unsuccessful. Inverting the soil and mixing it with uncontaminated sand—creating an abrasive effect which would release any dioxin bound to soil particles and leave it available for degradation—was also tried. "But the results are most disappointing," says La Porte.

Although Senator Noé is still faced with the problem of dealing with most of Zone A's high levels of dioxin, the employees of the Seveso Special Office—of which he is in charge—have been successful in edging Seveso back to normality. The decontamination procedures for the less affected Zones B and R involving removal of top soil and washing down buildings, particularly schools, has been completed. The dioxin concentrations in these areas are now very low, low enough for Senator Noé to be able to allow some farmers to begin planting, but only cereal crops in two areas of Zone R; root crops are, as yet, out of the question.

Noé has received official permission from the Ministry of Health in Rome for this limited project. As far as he is concerned it could not have come soon enough. Roche and Givaudan have also been eager to see the agricultural activity of the area return to normal as soon as possible. Farmers, too, have also been pushing for the restrictions to be lifted. But the international scientists brought in to advise Noé on the procedures to be adopted at Seveso have been guarded in their response. The scientists recognized the political importance of reopening the area to cultivation, but they refused to be hurried. Before they would agree to the procedure, they required assurances that it entailed no risk to the population. Italian scientists at Rome's Ministry of Health are confident that they can now give this assurance and they have advised Noé accordingly. The reason for their decision is twofold. Firstly the scientists claim that there is no evidence that dioxin is taken up by cereal crops and deposited in the seed grain. Secondly, they point out that dioxin concentrations in the soil to be cultivated are extremely low.

For two weeks in October 1980 even this limited project seemed in jeopardy when 150 sheep died in Zone R. The sheep had been transferred to Seveso for new pasture. By chance, the animals, unfed while in transit, had managed to leave their pasture and enter a field in Zone R where they consumed large quantities of fresh grass. The outcome was predictable. Unable to cope with the rapid fermentation in their rumen and the consequent release of gas, the animals suffered from bloat. Within 24 hours 150 were dead. The deaths led to the predictable supposition that dioxin might be the cause. But scientists at Milan's Mario Negri Institute rapidly established the correct cause of death as bloat.[43] Noé was relieved; his replanting scheme could continue.

Attempts to return to normal have been made in other areas besides that of agriculture. A problem confronting the Italian authorities in the wake of the evacuation of residents from Zone A was the question of what to do with their contaminated homes. Should they be demolished? Or could they be decontaminated? The view of the Cimmino Commission, a scientific commission of the Ministry of Health, chaired by Professor Aldo Cimmino, was unequivocal. Everything in Zone A would have to be destroyed.[44] The politicians in the Lombardy Regional Government disagreed. Dr. Vittorio Rivolta, the Region's Minister of Health, insisted that a detailed study be made of the situation before a final decision was reached.[45] Professor Augusto Giovanardi, a former director of the Milan Institute of Hygiene, was asked to make recommendations. Giovanardi instituted what later became known as the Seveso Decontamination Committee. Giovanardi's group worked swiftly. Less than two weeks after being asked to carry out its study, the Decontamination Commission presented its recommendations to the Lombardy Region's Health Council. Most of the bulidings in Zone A need not be demolished, the committee claimed, but could be decontaminated. Vegetables in the area, however, should be uprooted and burned. Topsoil could be left undisturbed in uncultivated rural areas, but should be removed from residential areas and incinerated.[46] The same procedures should also be applied in the less contaminated area, designated Zone B. The Decontamination Commission's proposals were accepted. Dr. Rivolta's pragmatism had paid off. The buildings at least were to be saved.

To implement the program the Decontamination Commission divided Zone A into seven sections (as shown in the figure). Areas A_1–A_3, those nearest the reactor, are the most seriously contaminated and contain about 80% of the discharged dioxin. Concentrations of the chemical ranged from

30 to 21,212 $\mu g/m^2$ (micrograms per square meter). A_4 and A_5 are areas of "medium contamination"—dioxin concentrations of 15–1043 $\mu g/m^2$. A_6 and A_7, where dioxin levels ranged from 15 to 475 $\mu g/m^2$, were the least affected portions of Zone A and housed 80% of the residents evacuated from the area by the local authority. Consequently it has been sections A_6 and A_7 which received most attention by the authorities.[21]

By request of the Italian authorities, Givaudan, the owner of the ICMESA chemical plant, was asked to decontaminate Zone A and to make it habitable again (Zone B was the responsibility of Italian officials). The dioxin level adjudged "safe" by the authorities in Zone A was 5.0 $\mu g/m^2$ for top soil and 0.75 $\mu g/m^2$ for all external hard surfaces such as walls and roofs, and 0.01 $\mu g/m^2$ for surfaces indoors.[46] By the end of June 1977 buildings in Zones A_6 and A_7 had been treated by a process of vacuum cleaning and washing with solvents. A complete layer of top soil and all vegetation had been stripped away and deposited in sections A_4 and A_5. Pavements and roads were the last areas to be cleared and when these had been completed, Givaudan claimed that the area was safe for reoccupation. Measurements of dioxin levels indicated that the program had reduced the concentration of the chemical to the target levels set by the Lombardy authorities. The 250-strong workforce employed by Givaudan carried out the work of decontamination with a thoroughness which was typically Swiss. But the work had not all been plain sailing. Givaudan and its parent company Roche claimed that it had been difficult work which had been carried out in the open, in very bad weather conditions, and at temperatures far below freezing. The job, the companies claimed, had also been made more difficult "by safety regulations that were not in all respects relevant to the actual situation on the site." However, they added that differences of opinion on this issue were of "no real significance."[47]

Of more consequence were the decisions concerning the best means of disposing of the contaminated top soil and vegetation from Zone A. Members of the Decontamination Commission had differing opinions about the methods to be applied to degrade the dioxin present. Incineration was the approach favored on theoretical grounds by some members of the Commission, but the view was by no means universal. Other scientists proposed trying degradation by the use of ultraviolet lamps, or removing the dioxin from soil by solvents followed by degradation using steam.[21]

Incineration was the only practical proposal but there was vehement opposition to it from the Seveso residents. They opposed the Commission's suggestion that an incinerator be constructed on the site of the ICMESA chemical plant. The residents had had enough. If the incinerator had to be built, they argued that it should be erected somewhere else. Knowing that incineration of the soil on the scale proposed would not only be an enormous undertaking without precedent, the residents claimed that they were to be the "guinea pigs" of the Lombardy authorities. In May 1977 they mounted a demonstration in Milan to protest against the incinerator's construction. The protests had the desired result; the incinerator was never built. Four and a half years after the accident the contaminated soil at Seveso is having to be placed in Senator Noé's sealed concrete bunker in the northwest corner of Zone A. There is nowhere else for it to go.

It is most unlikely that the more contaminated regions of Zone A will become habitable in the near future, if at all. Residents in the southern sections A_6 and A_7 returned to their homes in late 1977 with instructions from the Italian authorities that they were not to grow any vegetables and that for their own safety they should observe the strictest personal hygiene. The residents have been compensated in full by Givaudan for any losses incurred as a result of the reactor incident. Indeed Givaudan and Roche have settled some 6000 civil claims for damages at an estimated cost of 19.7 billion lira.[48] Lawyers for the companies recommended prompt settlement of claims to avoid higher bills in the future if the criminal court case—being brought by the Italian Judiciary against ICMESA, Givaudan, and Roche for operating a reactor which was unsafe—goes against them.[5]

The case has been in preparation ever since the reactor at ICMESA was sealed off on the orders of a local magistrate. Investigations into the cause of the accident at the ICMESA plant have been carried out on behalf of the Judiciary by Italian chemists and engineers, but without success. Givaudan has attempted to find an explanation but it, too, failed to come up with the answer. For four years the company attempted to simulate the accident under laboratory conditions which occurred on 10 July 1976. For this purpose the company constructed a smaller version of the ICMESA reactor with the same ingredients in the vessel and under the same conditions which prevailed on 10 July. The company tried to get the reactor to overheat, but to no avail. The reactor at Seveso was left at a temperature of 158°C when it was shut down. Six and a half hours later it had overheated sufficiently to cause an increase of pressure in the vessel. The pressure blew the safety valve venting the reactor contents over the residential area nearby.

During the manufacture of 2,4,5-trichlorophenol, dioxin formation is inevitable. The temperature at which the process is carried out depends on the solvent used in the mixture. Givaudan used ethylene glycol and carried out the process at 170–180°C. Any rise in temperature in the reactor will result in more dioxin being produced. But the critical temperature is 230°C; above this, conditions in the reactor are such that the reaction becomes exothermic and generates its own heat. The temperature of the whole mixture then rises rapidly. In a closed reactor, the rising temperature leads to a pressure increase, eventual rupture of the reactor, or the blowing of a pressure safety disk, as happened at Seveso.

The accident would not have happened had the technicians at the

ICMESA plant followed the agreed procedure of flooding the reactor with water when the process was completed. For some reason they had failed to do this. Givaudan's technical manager Dr. Jorg Sambeth cannot explain why this happened. Sambeth authorized the original procedure for the manufacture of trichlorophenol at ICMESA. He admits that water should have been added to the reactor but insists that this always was standard operating practice.[42] Even though ICMESA technicians deviated from normal practice on this occasion, research into the cause of the accident has suggested that there was nothing wrong with the chemistry of the process. The ingredients in the reactor were present in the correct proportions and by themselves could not have caused the temperature of the reactor to rise. Frustrated at not being able to identify the cause of the accident, Sambeth decided to investigate the possibility that some physical factor, such as the transfer of heat along the side of the reactor, could have been responsible. Professor T. G. Theofanous of the Department of Nuclear Engineering and Chemical Engineering at Purdue University, Indiana, was approached and asked to undertake the investigation.

Theofanous agreed to look into the matter without delay. An answer was forthcoming in a matter of months. Theofanous claims that heat transfer could well be the cause of the ICMESA accident.[49] The reactor can be visualized as being composed of two parts. The lower portion contains the reactor ingredients; the upper portion is simply the reactor shell. To heat the vessel superheated steam was passed through pipes attached to the outer walls. This heat was absorbed by the reactor contents in the lower portion of the reactor but in the upper section this did not occur; here the vessel walls were at a much higher temperature than the reactor contents. Until now it has been assumed that the heat from the upper section of a reactor would distribute itself evenly over the rest of the vessel and contents. Theofanous says this is not so. His research has shown that at the interface between the hot upper portion of the reactor and the surface of the vessel contents sufficient heat is available to raise the temperature of the contents above 230°C and hence start the exothermic reaction. Once started, the exothermic reaction would gradually move downward heating the rest of the contents in the process.[49]

This explanation is consistent with the state of the ICMESA reactor after the accident. Only the upper portion of the vessel's contents had overheated, the lower portion had not reached the critical temperature when the pressure blew the safety valve.

Between them, Theofanous and Sambeth have provided valuable information about the fallability of chemical reactors. Their findings will not pass unnoticed by the chemical industry; as a result, many companies will be forced to review their operating procedures.

Whether this information will influence the Italian criminal court case against the three companies remains to be seen. Sambeth has always argued that the trichlorophenol reactor at Seveso was designed after due account had been taken of previous accidents in trichlorophenol manufacturing processes. All previous accidents had occurred not after the process had been completed—as happened at Seveso—but during manufacture. The accident at Seveso was without precedent, says Sambeth.[42] Italian officials and scientists, however, are unlikely to find such protestations of primary importance in the light of the events at Seveso, and their grim consequences.

Equally grim is the toll of terrorist attacks on individuals associated with ICMESA, Givaudan and Roche, and with Seveso's Italian officialdom. The car of Givaudan's managing director, Guy Waldvogel, was blown up in Geneva. In Milan, Roche's offices were bombed, as was the house of the company's chief engineer Rudolf Rupp, who was sent to Seveso to assist with the decontamination program. Dr. Guiseppi Ghetti, the medical officer for Seveso and Meda, and one of the first Italian officials to know the extent of the pollution at Seveso, was attacked in his office in May 1977. Ghetti was shot five times, once through the shoulder and four times in the legs, one bullet shattering a kneecap. Paolo Paoletti, director of production at ICMESA, was the one individual likely to be most knowledgeable about operating conditions at the chemical plant; in February 1980 Paoletti was shot and killed by an extreme left-wing terrorist group.

Paoletti's death was a grim reminder of the situation which prevailed in Italy. At Seveso feelings have run high for some time. The residents of the region, angered by the reactor accident and the release of dioxin, tend now to vent their frustration on the Italian officials responsible for cleaning up the mess. In the past, part of the problem was the lack of continuity in the health and decontamination programs for the area. Heads have rolled on more than one occasion as chairmen of committees supervising this work have been replaced. Dr. Vittorio Rivolta, Minister of Health for the Lombardy Regional Government in 1976 and now the Region's Minister for the Environment, appointed two scientists to supervise work at Seveso. Professor Gaetano Fara and Professor Augusto Giovanardi were asked in August 1976 to head the Health and Epidemiology Commission and the Decontamination Commission, respectively. In June 1977 both commissions were dissolved, Fara and Givanardi lost their jobs and were replaced by a political appointee, Antonio Spallino, the Mayor of Como.

But Spallino's appointment as supervisor of the Lombardy Region's Special Office for Seveso did little to improve the flow of information to the residents. Official reports of developments at Seveso were rare. More often than not residents had to rely on newspaper reports—not always accurate—as a source of information. Spallino, often referred to as being aloof and unapproachable, was himself replaced in November 1979 by a former Senator of Milan, Luigi Noé. At the time of writing, Noé has been in charge of events at Seveso for over a year. Anxious to keep up the momentum of work, Noé, with the support of international scientists, has been instrumental in establishing a long-term health program for the Seveso residents. His plan for returning the area to some form of normality by hastening the decontamination program has also been well received by the residents and by scientists involved with the dioxin program. Noé's was a shrewd appointment; approachable, with experience in the world of politics, he is, in the opinion of many people, more capable than most of dealing with a problem which refuses to go away.

Roche and Givaudan have their problems too. The chemical industry has reacted unfavorably to the accident at Seveso. It should never have happened, the industry argues, and the two Swiss companies have been made aware that chemical manufacturers feel that their reputation has been tarnished by the accident. It remains to be seen whether Roche and Givaudan will recoup some of their lost prestige by their work to establish what went wrong at their ICMESA chemical plant. But the companies, or their insurers, will never recover the 67.7 billion lira paid in compensation to individuals, the Lombardy Region, and the Italian government for the damage and inconvenience caused by the release of dioxin at Seveso. A payment of this scale makes the companies wince. But money alone will not remove the scars caused by the accident. The pregnant women of Seveso can hardly be expected to forget the anxiety they experienced over dioxins potential threat to their unborn children. Roche and Givaudan will not be allowed to forget the episode for a while yet, for the lessons of Seveso will feature promiently in the criminal court case still to be brought against the companies by the Italian Judiciary.

References

1. Hay, A. Toxic cloud over Seveso, *Nature (London)* 262, 636–638 (1976).
2. Letter from ICMESA management to Seveso Health Officer, 12 July 1976.
3. Evento Icmesa. Cronistoria, Guinta Regionale, Assessorate alla Sanita, Milano.
4. Reggiani, G. Personal interview (24 June 1977), Margerison, T., Wallace, M., and Hallenstein, D. *The Superpoison*, Macmillan, London (1979), p. 81.
5. Hay, A. Italian Commission covered up Seveso delays, *Nature (London)* 277, 588–589 (1979).
6. Margerison, T., Wallace, M., and Hallenstein, D. *The Superpoison*, Macmillan, London (1979).
7. Reggiani, G. Personal interview (16 April 1978); Margerison, T., Wallace, M., and Hallenstein, D. *The Superpoison*, Macmillan, London (1979), p. 103.
8. Waldvogel, G. Letter to Health Officer of Seveso and Meda, 23 July 1976.
9. Hay, A. Tetrachlorodibenzo-*p*-dioxin release at Seveso, *Disasters* 1, 289–308 (1977).
10. Whiteside, T. *The Pendulum and the Toxic Cloud*, Yale University Press, New Haven and London (1979).

11. Commissione Parliamentare di Inchiesta sulla fuga di Sostanze tossiche ovvenuta il 10 Luglio 1976 nello stabillmento Icmesa e suit rischi potenziali per la salute e per l'ambiente derivanti da attitita indistriali, Camera de Deputati, Doc. XXIII, No (25 July 1978). Available in English translation from U.K. Health and Safety Executive.
12. Hay, A. What caused the Seveso explosion? *Nature (London)* **273**, 582–583 (1978).
13. Press report of Italian Parliamentary Commission of Inquiry (July 1978).
14. *Thrice 25 years. Fragments from the Roche Story*, Roche-Zeitung (1971).
15. Anonymous. *La Notte*, 13 August 1976.
16. Anonymous. *Corriere Della Sera*, 12 August 1976.
17. Anonymous. *Il Tempo*, 13 August 1976.
18. Anonymous. Twelve months after Seveso, *Nature (London)* **268**, 90 (1977).
19. Commissione Parliamentare, Camera de Deputati, Doc. XXIII, No. 6 (25 July 1978), p. 128.
20. Diaguardi, N. Critical considerations three years after the accidental pollution at Seveso, Instituto di Clinica Medica II Della Universita di Milano (1979).
21. Hay, A. Seveso solicitude, *Nature (London)* **267**, 384–385 (1977).
22. Fara, G. Paper presented at Workshop on Impact of Chlorinated Dioxins and Related Compounds on the Environment, Rome (22–24 October 1980).
23. De Carli, L. Cytogenetic investigations on the Seveso population exposed to TCDD, International Workshop on Plans for Clinical and Epidemiological Follow-Up after Area-Wide Chemical Contamination, National Academy of Sciences/National Research Council (17–19 March 1980).
24. Puccinelli, V. Working paper for chloracne panel, Commission of the European Communities, Doc. No. 1320/77, Milan (11–12 July 1977).
25. Noe, L. Risk evaluation in industrial activity—the Seveso case, 3rd Advanced Seminar on Risk and Safety Assessment in Industrial Activities, ISPA (16–20 June 1980).
26. Noé, L. Personal interview, 20 October 1980.
27. Margerison, T., Wallace, M., and Hallenstein, D. *The Superpoison*, Macmillan, London (1979), p. 129.
28. Margerison, T., Wallace, M., and Hallenstein, D. *The Superpoison*, Macmillan, London (1979), p. 132.
29. Margerison, T., Wallace, M., and Hallenstein, D. *The Superpoison*, Macmillan, London (1979), p. 137.
30. Gropp, A. Letter to Assessore V. Rivolta, Lombardy Regional Health Authority, 2 February 1977, and quoted in Reference 21.
31. Rehder, H., Sanchioni, L., Cefis, F., and Gropp, A. Pathologischembryologische Untersuchungen an Abortusfallen in Zusammenhang mit dem Seveso-Ungluck, *Schweizerische Medizinische Wochenschrift* **108**, 1617–1625 (1978).
32. Fara, G. M., and Marubini E. Monitoring birth defects: an Italian project, the 3rd conference of the European Teratology Society, Helsinki, *Teratology* **10**, 309 (3–6 June 1974).
33. Reggiani, G. Personal letter, 1 March 1979.
34. Tuchmann-Duplessis, H. Pollution dc l'environnement et descendance a propos de l'accident de Seveso, *Medicine et Hygiene* **36**, 1758–1766 (1978).
35. Commissione Parliamentare, Camera di Deputati, Doc. XXIII, No. 6, (25 July 1978), p. 131.
36. Klingberg, M. A. Summary and conclusions derived from the papers Human health effects from accidental release of TCDD at Seveso, and Mortality and birth defects from 1976 to 1979 among the population of the Seveso area, presented at workshop on Impact of Chlorinated dioxins and Related Compounds on the Environmnent, Istituto Superiore de Sanita, Rome (22–24 October 1980).

37. Spontaneous and induced abortion. World Health Organization Technical Report Series No. 461 (1970), pp. 20-26.
38. Bisanti, L., Bonetti, F., Caramachi, F., Del Corno, G., Favaretti, C., Giambelluca, S. E., Marni, E., Montesarchio, E., Puccinelli, V., Remotti, G., Volpata, C., Zambrelli, E., and Fara, G. M. Experiences from the accident of Seveso, *Acta Morphologica Academiae Scientiarum Hungaricae* **28**, 139-157 (1980).
39. Commissione Parliamentare, Camera de Deputati, Doc. XXIII, No. 6 (25 July 1978), p. 169.
40. La Porta, V. Personal interview, 20 October 1980.
41. Di Domenico, A., Silano, V., Viviano, G., and Zapponi, G. Accidental release of 2,3,7,8-tetrachlorodibenzo-*p*-dioxin (TCDD) at Seveso (Italy): V. Environmental persistence (half-life) of TCDD in soil, Istituto Superiore di Sanita (26 January 1980).
42. Hay, A. Seveso: the crucial question of reactor safety, *Nature (London)* **281**, 521 (1980).
43. Walgate, R. Bloat, not poison. *Nature (London)* **287**, 6 (1980).
44. Anonymous. *Corriere della Sera*, 12 August 1976.
45. Anonymous. *La Notte*, 14 August 1976.
46. Giovarnardi, A. Decontamination of the dioxin-contaminated areas, Guinta Regionale della Lombardia Assessarato Alla Sanita (4 December 1976).
47. Anonymous. Seveso Survey No. 4: Roche Nachrichten, June 1977.
48. Anonymous. Seveso Survey No. 9: Roche Nachrichten, 1980/82.
49. Theofanous, T. G. The cause of the Seveso accident: a physicochemical mechanism, Dechema Colloquium on the Seveso accident, Frankfurt, West Germany (November 1980).

10

Love Canal

Disposal of dioxin-contaminated soil at Seveso presents a problem which remains as yet unresolved. Soil containing 250 g of dioxin lies untouched in the contaminated area designated "Zone A" as the authorities debate how to deal with the situation. Many now seem to favor the only real option open to them—burial in a deep landfill in the middle of the zone.

But landfills may present their own problems, as the residents of the city of Niagara Falls, New York, know all too well, in particular the members of 1000 families living in an area previously used for the dumping of chemical wastes by the Hooker Chemical Company. The site is known as Love Canal.

Originally designed in 1896 by William Love as a navigable power canal between the upper and lower Niagara Rivers, it was to have been so engineered as to provide water power for local industry. Considered by many to be a good investment, the project fell on hard times when severe economic recession made funding impossible. The death knell for the project was the demonstration by Louis Tesla in 1910 of a means of transmitting electrical power over long distances by means of an alternating current. Industry had no longer to be sited close to a source of power, which could now be delivered to its door.

The pace of scientific change had overtaken Love's grandiose scheme. Excavation work had started before the project was finally cancelled and a section of the canal in the southeast corner of the city had already been dug. For over a decade the excavation lay untouched until the late 1920s when a new use was found for it. The chemical industry, growing at a furious pace, was desperate for a site to deposit its waste products. Love Canal was ideal for this. The site was already dug, and waste had only to be placed in it.

Hooker Chemicals began dumping waste in the Canal in about 1930 (an exact date has not been published). The waste included chlorinated hydrocarbons, residues from chemical processes, and ash from incinerators. According to Hooker's records, in the ten-year period ending in 1952, the company had dumped some 21,800 tons of material into the canal bed.[1] At least 200 separate chemicals have been identified in this waste, including benzene, toluene, trichloroethylene, dibromomethane, benzaldehyde, together with most of the chlorinated benzenes and toluenes used in the chemical industry.[1] Eleven of the chemicals are known or suspected carcinogens.[2]

Some of the waste chemicals identified are the residues remaining from the production of herbicides. Herbicides and pesticides represent extremely profitable business for many companies. Hooker recognized the potential of herbicides and was one of the first producers of 2,4,5-T (2,4,5-trichlorophenoxyacetic acid). One stage in the manufacture of the herbicide involves the production of 2,4,5-trichlorophenol; it is at this point that dioxin is generated and, in a chemical reactor, it settles out in the residue at the bottom of the vessel. At Hooker, this residue was treated like any other waste and was dumped in the canal. A senior U.S. Environmental Protection Agency (EPA) Official has admitted that, on the basis of the estimated, 2,4,5-trichlorophenol present in the canal site, there could well be as much as 5 kg of dioxin there in addition.[3] This would mean, in other words, the site may have an estimated 20 times as much dioxin as was discharged over Seveso. It is unlikely that the true figure will ever be known.

The problem posed by Love Canal would have been serious enough had it been situated in a secluded spot; but it is, in fact, set in the middle of a residential area with houses flanking it on both sides.[4] Such siting would be inconceivable today, but in 1953 the city of Niagara Falls was very interested in the construction of houses and the precincts of Love Canal seemed a suitable location. The authorities asked for and were sold the site for a nominal $1 by Hooker. With the transaction completed the canal was filled in and house building began adjacent to the site. Eventually, 99 houses and a primary school were built directly alongside the canal.[5]

According to Dr. David Axelrod, Health Commissioner for New York State, worries about potential hazards posed by the site began to surface in the Spring of 1978.[1] Occasional complaints from residents about malodorous substances had been noted prior to this but no action was taken. However, there was an unusually heavy fall of snow in the early months of 1978. In the ensuing thaw many of the chemicals, previously well buried, were brought to the surface. The localized flooding which occurred carried chemical waste into many basements in the houses on the canal edge. The smell was appalling and, following complaints from residents, an investigation was ordered. It soon became clear that the complaints were justified. Vapors identified as emanating from toxic chemicals were percolating from the basements into the rest of the homes.[5]

The toxicity of the chemicals now deposited in both homes and school was such that action had to be taken, and events moved rapidly after this, as the following chronology makes clear. On 15 April 1978, the Department of Health suggested that Love Canal residents faced a health hazard. By 13 May EPA scientists had confirmed this hazard in their analysis of the toxic vapors. Residents were told the results of these investigations on 19 May. On 21 May short- and long-term medical studies to find any health problems were instituted. Further investigations by State Officials and discussions with eminent scientists confirmed the gravity of the situation, and on 2 August the Governor of New York State, Hugh Carey, declared that a state of emergency existed at the Love Canal site. Steps were taken to close the school and evacuate pregnant women and children under two years of age. On 17 August President Carter approved financial aid for the Love Canal area. Two days later plans were made to evacuate all 236 families living in the vicinity of the canal. Within weeks the residents were found alternative accommodation. An eight-foot-high chain link fence was erected around the site.

The above account places State officials in a favorable light and suggests that action to protect the residents was instituted promptly. This was not always the case, according to Michael Brown, in his book *Laying Waste*.[6] As a reporter formerly employed on the *Niagara Gazette*, Brown followed every twist and turn in the developing story at Love Canal. Brown documents instances where Niagara Falls city officials and Hooker Chemical's management had had private discussions on the seriousness of the situation at the canal. These took place well before there was any public announcement of the danger. But Brown claims that before this, in 1976, the New York State Department of Environmental Conservation carried out its own tests to check the area and found to its disquiet that toxic chemicals were flowing from the canal into adjoining sewers.

Prior to 1978, the health problems among the Love Canal residents had not been related to pollution by toxic wastes. No survey had been made to assess the problem; indeed there appeared to be no reason for such a study. Even today few details are available about the health of the residents of the Love Canal area. Health surveys of the residents in the immediate vicinity of the canal were enlarged to include previous residents in the area and those in neighboring streets. By 18 August 1978 some 2800 residents had given blood for routine tests to determine their general state of health. In the ensuing months, residents living within a four-block radius of the landfill were prevailed upon to answer a 29-page questionnaire designed by Department of Health epidemiologists. The questionnaire sought detailed information on "present and past state of health, family history, occupations and length of stay in the area."[5]

The small numbers involved in this survey makes interpretation of the results difficult and conclusions drawn from the data must also be treated with caution. After detailed investigation some conclusions have been reached. The incidence of babies born with low birth weights ($5\frac{1}{2}$ lb or less) in the contaminated areas was 13.9%, almost twice the 7.27% incidence in the noncontaminated area (the incidence in the State of New York as a whole is 6.97%).[5] The evidence also suggests that among residents in one of the more severely contaminated streets there has been a slight-to-moderate increase in spontaneous abortions. A higher incidence of malformed children has also been noted for one of the more severely contaminated areas. But according to one of the investigators involved, the increase is not statistically significant.[5,7] The small numbers of births involved make analysis of this kind of data extremely complex.

More medical problems may surface in future investigations. For any that do arise it will prove exceedingly difficult to find a single causative agent. Most of the chemicals identified in the landfill are toxic and undoubtedly harmful; dioxin is just one of the many in this category.

A suggestion that this cocktail of chemicals had caused a serious medical problem occurred in 1980. Events took a dramatic turn on 17 May 1980 following an announcement by the Environmental Protection Agency that a pilot study had revealed an increased incidence of chromosome aberrations in the blood of Love Canal residents. The study, performed by Dr. Dante Picciano, a genetic toxicologist at the Biogenics Corporation of Houston, had found that of the 36 residents examined, 11 had a significant number of aberrations, including breaks, marker chromosomes, and ring chromosomes. Eight of the eleven had marked damage in the form of "supernumary acentric chromosomes," a finding Dante says normally occurs in one out of 100 individuals.[8]

According to an EPA official the results of Dante's study had been leaked to a newspaper. This had put the Agency in a difficult position. The EPA's intention all along had been to study the Picciano results in detail, discuss their implications and consider whether action was required. This approach had now been overtaken by events and the Agency had to respond. The results of Picciano's study were announced publicly to an audience of Love Canal residents by the EPA's deputy director, Barbara Blum.

In her statement, Blum advocated caution in the interpretation of Picciano's findings, pointing out that they were only the results of a preliminary study. Blum's qualification was timely. Four days after the chromosome results were announced by the EPA, a panel of scientists from the Department of Health and Human Services (DHHS) denounced the study as worthless. The DHHS panel had attempted to obtain Picciano's original research findings (important for a proper assessment, as the original scientific data could be subjected to statistical treatment in a manner which was inappropriate for a particular set of results. Furthermore, there could be mistakes in the statistical analysis). Picciano refused to hand over his original data and the panel had to be satisfied with the report of his findings which he submitted to the EPA. Commenting on this report the DHHS panel stated that the data did not support the assertion that there was an increase in chromosome damage in the Love Canal population. The "reported increases in chromosome breaks and marker chromosomes were not statistically significant" the panel claimed.[2] But the most serious fault the panel stated was Picciano's failure to compare the results with those from a matching unexposed control group. Without this comparison the panel concluded that the study was of little value. Six days later, on 27 May, the EPA convened its own panel to review the study. This tme the original data were provided. Its provision had little impact on the review panel, which issued an equally crushing opinion. The EPA panel concluded that there was virtually no value in the Picciano findings.[2] A second, larger properly controlled investigation on chromosome damage was therefore planned for later in the year.[9]

The reviews were issued too late to prevent President Carter from signing an executive order on 21 May declaring a State of Emergency at Love Canal, for the second time in two years. Whether the reviews would have influenced his decision to sign is a moot point. There was already sufficient evidence to suggest that chemicals had leached from the canal to contaminate an area far greater than that in its immediate vicinity.[6]

Following President Carter's declaration, orders were issued for the evacuation of an additional 710 families. Although some residents expressed reservations about leaving their homes, most were prevailed upon to move in the knowledge that the cost of living elsewhere would be met for one year at least by public funds. Just who pays is another matter. The State of New York has insisted that it is the Federal Government's responsibility. After all, it argues, it was the government which ordered the evacuation of the residents and the State should be reimbursed for the cost of buying abandoned homes at Love Canal. The Federal Government has claimed that it lacks the authority to pay up.[2]

The full cost of monitoring the health of the Love Canal residents, rehousing the evacuees, and rendering the landfill safe will be decided by litigation. As a result of its dumping practices, Hooker Chemicals is being sued both by individuals and by government agencies. Private litigants are demanding some $14 billion from the company in compensation. The government agencies' claims are more modest. Even though officials in the EPA acknowledge that the dumping procedures adopted by Hooker were considered to be the best engineering practice at the time, the company is still expected to pay its way. The Justice Department and the EPA are suing Hooker for $125 million for the cleaning-up operation. New York is claiming $635 million from Occidental Petroleum Corporation, Hooker's parent company, for the "public health hazard and environmental disaster" at Love Canal.

The suits against Hooker are punitive. The actions, if allowed, will deal a serious financial blow to the company.[9] The EPA is well aware of this but in taking this action against Hooker, the Agency is serving notice on other chemical companies that it intends to take action where necessary to control the movement or disposal of toxic wastes. The warning is not likely to go unheeded.

With the number of landfills containing toxic wastes in the United States estimated to be in the tens of thousands, the EPA clearly has a major problem on its hands. Hooker itself has three other waste disposal sites in addition to that at Love Canal.[10] The largest of the dumps, referred to as the Hyde Park site, is also in the middle of a residential area and located near a university campus and a power station. Potentially more serious than the Love Canal site, Hyde Park alone contains an estimated 120 kg of dioxin—the largest single dump of the chemical known. Many other chemicals are in the Hyde Park landfill and the EPA intends to make this area the subject of a major investigation.

Although Hooker is something of a test case in the EPA's campaign against poor waste management practice by other large chemical enterprises, the company will undoubtedly survive the litigation. Had Hooker not sold Love Canal to the City of Niagara Falls for the nominal $1, but simply handed the land over, the company could not have been prosecuted for malpractice; the city itself would probably have been liable. For this reason, there are those who believe that a decision by the Supreme Court may be necessary to resolve the issue. Whoever foots the final bill—and funds may well come from the U.S. Treasury—Love Canal, like Seveso, will have left its scars on a great many people. Ominously, the EPA's former Administrator Douglas Costle has predicted that even this is only the tip of the iceberg. Many more situations will turn up he says "which will shock our nation."[11]

References

1. Axelrod, D. Chlorinated hydrocarbons (US Love Canal): case studies of selected areawide environmental exposures, presented at the National Academy of Sciences (NAS) Workshop on Plans for Clinical and Epidemiological Follow-Up after Area-Wide Chemical Contamination, Washington D.C. (17 March 1980).
2. Raloff, J. Disaster on 99th Street, *New Scientist* **86,** 298–300 (1980).
3. de Rosiers, P. Personal communication, March 1980.
4. Love Canal: Public Health Time Bomb, A special report to the Governor and Legislature, September 1978 prepared by the Office of Public Health for the State of New York and the Governor's Love Canal Interagency Task Force.
5. Axelrod, D. Adverse effects on target sites: reproductive injury, Presented at NAS workshop, Washington D.C. (17 March 1980).

6. Brown, M. *Laying waste*, Pantheon Books, New York (1980).
7. Vianna, N. J. Adverse pregnancy outcomes—potential endpoints of human toxicity in the Love Canal (preliminary results), in *Human Embryonic and Fetal Death*, ed. I. H. Porter, and E. B. Hook, Academic Press, New York (1980).
8. Pilot cytogenetic study of the residents of Love Canal, New York, Biogenics Corporation, Houston, Texas (14 May 1980).
9. Hooker hunts PR man to boost sagging image, *Chemical Week* (June 6, 1979), p. 18.
10. Discovery of dioxin at Montague could impede accord on Hooker wastes, *Chemical Week* (June 6, 1979), p. 18.
11. Blumenthal, R. *The New York Times*, 30 June 1980.

11

Where Do We Go from Here?

The history of 2,4,5-trichlorophenol and 2,4,5-T can with some justification be described as appalling. As the preceding chapters have shown, both chemicals have caused their share of suffering. Central to the problem is the presence in both of the contaminant dioxin. Workers exposed to high concentrations of dioxin in industry have invariably developed the skin disease chloracne. But this disfiguring condition appears to be only one part of their health problems. The evidence now accruing suggests that these men may be at greater risk of developing soft tissue cancer.

Most manufacturers of trichlorophenol have experienced accidents in which workers have been exposed to dioxin. Are there, then, any conclusions to be drawn from these incidents? And, if so, what needs to be done in the future to prevent a recurrence?

There are some obvious lessons for manufacturers of chemicals which have as poor a track record as 2,4,5-trichlorophenol. Communication between companies has been a problem in the past. Some companies have been open and helpful, others have been the reverse. The example was cited of the British company Coalite and Chemical Products Limited requesting information from the U.S.-based Monsanto company on the steps to be taken following an accident at Coalite in 1968 and involving 2,4,5-trichlorophenol with heavy dioxin contamination; Monsanto had had just such an accident in 1949. The American company clearly had useful information to offer but, according to reports, chose not to provide it. Coalite, in turn, despite the earlier apparent willingness to disclose information, has been equally guilty in this regard by refusing to provide details of its accident and the report of it to Givaudan, the Swiss owners of the trichlorophenol reactor at Seveso.

Givaudan believes it now knows what caused the reactor to overheat and to discharge dioxin over a populated area at Seveso.[1] The company's own research in this area has been published as has that from research being carried out by others on its behalf. The information, now available, will be of inestimable value to chemical manufacturers. In its earlier more communicative mood, Coalite too published details about its own trichlorophenol accident in 1968, and explained why temperature was such a critical factor in the production of the chlorinated phenol.[2] However, Coalite has not released details which point conclusively to the cause of the 1968 accident.

Evidence presented by the company to the Factory Inspector investigating the accident, and also given at the coroner's inquest on Eric Burrows, the chemist killed when the reactor ruptured, points to overheating as the primary cause.[3-6] The definitive report of the Coalite accident, however, is in the possession of the U.K. Health and Safety Executive (HSE). Twelve years after the accident at Coalite, and a full four years after the company ceased production of 2,4,5-trichlorophenol, the HSE was still refusing to release this information.[7] The grounds for the HSE's decision was that it would be a breach of confidence were it to publish its report of the accident. There are some in the Executive who disagree with this position but the legal department advised against publication and it is the HSE's lawyers who have won the day. Without this information it is impossible to compare the Coalite accident with ones which occurred in other trichlorophenol plants, a comparison which, it could be argued, would make the process that much safer for other manufacturers. Fortunately Givaudan has had no such qualms about publishing the results of its own investigations into the cause of the Seveso accident.

Clearly the company's actions were not motivated purely by altruism. Establishing that its reactor at Seveso was built to the best engineering standards of the day is a vital part of Givaudan's defense case for the forthcoming trial in Italy which will deal with the accident. Givaudan will undoubtedly argue that the accident was unforeseen and that it was not attributable to negligence in the design of the reactor.[8] The company's own research and the results of studies which it has sponsored strengthen its claim. But it remains to be seen whether its plea is accepted by the court. By publishing its research results Givaudan is clearly hoping to win approval for its case. But it could be argued that this could just as easily have been done in court. Perhaps so. Nevertheless, the lessons and the probable cause of the Seveso accident are now available for scrutiny and public debate. This is as it should be. All companies should be encouraged to be equally forthcoming when discussing accidents of consequence to the public.

But the lessons to be drawn from Seveso go far beyond the arguments about the design features of trichlorophenol reactors and their modes of operation. The real tragedy of Seveso has been the effect of the accident on the residents of the town. For most of them the whole episode has been a nightmare. It began with the delay in evacuating people from the dioxin-contaminated zone. When they were eventually moved, they were housed in hotels in the neighborhood. To their dismay other hotel residents treated them like lepers shunning them because of their contact with a highly toxic chemical. At that time, the effect of exposure to dioxin on their health was unknown. Much of this information is now available. No deaths are directly attributable to the release of dioxin. Other medical problems were of a temporary nature and may have resolved in time. As for the children with chloracne many are now clear of the skin disease and no new cases have been recorded for some time, testimony to the efficiency of the decontamination and personal hygiene programs that were implemented.

It was the abortion issue at Seveso, however, which is likely to remain as the worst legacy of the dioxin accident. At least 100 women are reported to have sought abortions in the wake of the discharge of dioxin fearing the teratogenic properties of the chemical.[9] Any sympathy the women might have expected from local medical practitioners was, with a few exceptions, not forthcoming. Doctors proved frequently to be insensitive to the needs of their patients. They would not countenance abortions, and as a result many women were forced to seek the operation outside Italy. As it turned out, the risk to the developing fetus from the dioxin at Seveso was less than had been feared initially. But at the time neither the women nor their doctors knew this. All the more reason, therefore, why Italian officialdom might have done more to help. This episode was humiliating for all concerned and fortunately unlikely to be repeated with the new law permitting abortion on medical grounds for Italian women.

Seveso was a unique type of accident. But it would be unwise to assume that it might not be repeated elsewhere. It was dioxin this time, but on another occasion it might be another chemical equally as toxic. The evidence assembled in this book on the fallout—chemical and otherwise—from industrial accidents is hardly conducive to complacency. A more appropriate reaction might be "There but for the grace of God . . ." which probably explains why many government regulatory authorities have taken the Seveso accident to heart when analyzing the dangers for residents living close to any heavy industrial site.

There are dangers too for the workers in industry. Those employed in the manufacture of 2,4,5-trichlorophenol have had more than their share of problems. Poor manufacturing processes where industrial hugiene has been rudimentary have been the cause of many developing chloracne, a most persistent and disfiguring skin disease. As the cause of the skin complaint is known to be due to the presence of dioxin contaminants, there is little exclue for its occurring in future. Manufacturers should be prevailed upon to observe the strictest industrial hygiene in their chemical plants.

But skin problems, although serious, are not the most worrysome feature as far as industrial workers who are still engaged on trichlorophenol production are concerned. As more mortality studies of industrial workers exposed to dioxin are carried out there is a suggestion that deaths which can be linked with exposure to the chemical are not as numerous as was at first supposed.[10,11] But the picture is still far from clear. The incidence of soft tissue cancer in dioxin-exposed workers in industry is high.[12,13] It is also reported to be high in forestry workers in Sweden who have used 2,4,5-trichlorophenol.[14] But the authors of a study of Finnish workers using 2,4,5-T have not observed this same high incidence of soft tissue tumors.[15] These apparently conflicting reports only make more urgent the need for additional mortality studies on dioxin-exposed workers to resolve the issues and to enable an assessment to be made of the long-term risks of exposure to dioxin.

The Environmental Protection Agency (EPA) in Washington is still concerned about the threat to the individual from exposure to 2,4,5-T and its dioxin contaminant; the Agency is known to be worried about the implications of the Swedish findings and the suggested link between herbicide exposure and cancer. It is also known that the EPA is anxious for other studies to be undertaken to confirm or refute the evidence from Sweden.

The EPA is only one of a number of government institutions in the United States concerned about 2,4,5-T and "Agent Orange," the preparation of the herbicide used in Vietnam. The Veterans' Administration and the Department of Health and Human Services (the latter previously the Department of Health, Education and Welfare) are both investigating the effects on health of exposure to Agent Orange by former U.S. veterans of the Vietnam War. These studies will be wide ranging. They will concentrate on the incidence of cancer from among the veterans, their general physical condition, and the rate of malformation in their children. The design of these investigations is crucial; they should be able to provide a clear answer to the effects of exposure to Agent Orange. The expense of mounting these studies is considerable and it is most unlikely that funds will be forthcoming for future work if the current program should prove insufficient.

A further group of Vietnam veterans, also exposed to Agent Orange, is being studied in Vietnam itself. Vietnamese scientists claim that the herbicide could be the cause of a high incidence of birth defects in the children of Vietnamese war veterans.[16] These claims have yet to be subjected to the scrutiny of other scientists and will require validation by studies carried out on a larger scale. At present some U.S. scientists are not convinced by the Vietnamese claim. Others accept that exposure to Agent Orange has had its effect, not just on veterans but on the general population of Vietnam. A study by the American Association for the Advancement of Science (AAAS) noted an increase in the incidence of two birth defects—spina bifida and cleft palate—in children born to women who were resident in areas of Vietnam heavily sprayed with Agent Orange.[17]

Where Do We Go from Here? 241

Disagreements about the design and validity of studies carried out in Vietnam are at the center of the controversy. But these differences of opinion are not irreconcilable. If U.S. and Vietnamese scientists were convinced of the need to collaborate with one another to resolve the problem of experimental design, more information would be available to assess the risks—if any—posed by exposure to Agent Orange.

There is more than one dioxin, however. Agent Orange was contaminated by the most toxic of this class of chemicals—2,3,7,8-tetrachlorodibenzo-*p*-dioxin. But there are other dioxins which, although less toxic than the herbicide contaminant, are very much a subject of concern for scientists and government regulatory authorities. As the chapter on the chemistry of the dioxins (known technically as polychlorinated dibenzodioxins) makes clear there are many sources of these chemicals besides 2,4,5-trichlorophenol and 2,4,5-T. Industrial and municipal incinerators are known to discharge a variety of polychlorinated dibenzodioxins in their flue gases. It is not always known why this should occur. Much more research is required to investigate the possible risks posed by the situation as well as the nature of the substances in the incinerator which combine on heating to form polychlorinated dibenzodioxins. But these dioxins need not be a permanent feature of flue gases. The Dow Chemical Company has demonstrated that many of the dioxins can be removed from the flue gases by additional burners.[18]

Analytical techniques to detect polychlorinated dibenzodioxins are constantly being improved. The scope of this book has only allowed for brief reference to be made to some of the techniques available. The field has been reviewed admirably and in depth elsewhere; interested readers should refer to these sources for details.[19-21] As knowledge advances, it has become increasingly clear that investigations of the toxicology of dioxins will have to concentrate on the reasons for one type of dioxin being more toxic than another. It is still not clear how the most toxic of the dioxins—2,3,7,8-tetrachlorodibenzo-*p*-dioxin—acts to cause death. Its ultimate site of action in living tissues is unknown. The fate of dioxins in soil is also a subject requiring further study as there is as yet no measure of agreement between scientists about the environmental degradation of these chemicals.

Answers to these questions are urgently needed. It is doubtful whether they will be forthcoming before the Environmental Protection Agency decides the fate of 2,4,5-T. Some of those from within the EPA adhere to the view that the days of this herbicide are not necessarily numbered, as was first supposed. They argue that the EPA may well say that it will allow 2,4,5-T to remain in use provided its dioxin content is further reduced. This compromise will find favor with the chemical companies who manufacture the herbicide. Were the EPA to lose—as it has done in the past on this issue—it would be a considerable blow to the Agency's prestige. Were Dow

Chemical to lose the argument, the company would not let the matter rest but appeal against the decision and pursue it through the courts. The litigation costs would be considerable. A compromise is thus a real possibility.

Lowering the dioxin contamination of 2,4,5-T and improving industrial hygiene in chemical plants where the herbicide and its precursor 2,4,5-trichlorophenol are manufactured would please many scientists. Although alternative herbicides are available some scientists argue that they have not been as extensively researched as 2,4,5-T.

Some government review bodies in countries other than the United States have already reached the conclusion that 2,4,5-T is safe. The evidence suggests that more review bodies will do likewise, but while recommending the continued use of the herbicide they are also likely to suggest that workers using the chemical as well as those manufacturing 2,4,5-T and its precursor 2,4,5-trichlorophenol should receive routine health checks.

These recommendations formulated with the benefit of hindsight would undoubtedly improve the situation for users of 2,4,5-T in future. For those who have been exposed to the herbicide or its dioxin contaminant in the past, including the people of Vietnam, the veterans of the Vietnam War, the residents of Seveso, or workers in the chemical industry, no panacea is available. Only further investigations will give any indication of their health prospects in the long term. But even these answers are unlikely to be forthcoming in the immediate future. It is likely to be a long haul.

References

1. Theofanous, T. G. The cause of the Seveso accident. A physicochemical mechanism, Dechema Colloquium on the Seveso Accident, Frankfurt, West Germany (November 1980).
2. Milnes, M. H. Formation of 2,3,7,8-tetrachlorodibenzodioxin by thermal decomposition of sodium trichlorophenate, *Nature (London)* **232**, 395–396 (1971).
3. Payne, K. R. 2,4,5-trichlorophenol production process. Fine chemicals, evidence given at inquest on Eric Burrows, 26 April 1968
4. Anonymous. 2,4,5-trichlorophenol production process, process details with special reference to the reaction stage, CM/RER/26 (April 1968).
5. Ferguson, D. Notes on fatal accident to Eric Burrows, Sheffield South accident 1241/1968, explosion at Coalite Oils and Chemicals Ltd. Bolsover, Derbyshire (3 May 1968).
6. Gerrard, J. W. Testimony to H. M. Factory Inspectorate, Ministry of Labour (24 April 1968).
7. Hammer, J. D. G. (H.M. Chief Inspector of Factories). Personal letter, 18 November 1980.
8. Hay, A. What caused the Seveso explosion? *Nature (London)* **273**, 582–583 (1978).
9. Hay, A. Seveso solicitude. *Nature (London)* **267**, 384–385 (1977).
10. Zack, J., and Suskind, R. The mortality experience of workers exposed to tetrachlorodibenzodioxin in a trichlorophenol process accident, *Journal of Occupation Medicine* **22**, 11–14 (1980).
11. Zack, J., and Suskind, R. (to be published). Pr Newswire, St. Louis, Missouri (20 October 1980).

12. Honchar, P. A., and Halperin, W. E. 2,4,5-T, Trichlorophenol and soft tissue sarcoma, *Lancet* **I**, 268-269 (1981).
13. Cook, R. R. Dioxin, chloracne, and soft tissue sarcoma, *Lancet* **I**, 618-619 (1981).
14. Hardell, L., and Sandstrom, A. Case control study: soft tissue sarcomas and exposure to phenoxyacetic acids and chlorophenols. *British Journal of Cancer* **39**, 711-717 (1979).
15. Riihimaki, V., Asp, S., and Hernberg, S. Mortality of chlorinated phenoxyacid herbicide (2,4-D and 2,4,5-T) spraying personnel in Finland, an ongoing prospective follow-up study (unpublished).
16. Tung, T. T., Lang, T. D., and Van, D. D. Le probleme des effets mutagenes sur la premiere generation apres l'exposition aux herbicides (unpublished) (1979).
17. Senate Congressional Record S 13226-S 13233 (3 March 1972).
18. Crummett, W., Bumb, R. R., Lamparski, L. L., Mahle, N. H., Nestrick, T. J., and Whiting, L. Environmental chlorinated dioxins from combustion—the trace chemistries of fire hypothesis, presentation at Workshop on Impact of Chlorinated Dioxins and Related Compounds on the Environment, Rome (to be published by Pergamon Press) (22-24 October 1980).
19. Bovey, R. W., and Young, A. L. *The Science of 2,4,5-T and Associated Phenoxy Herbicides*, John Wiley & Sons, New York (1980).
20. Hass, J. R. and Fiesan, M. D. Qualitative and quantitative methods for dioxin analysis, *Annals of the New York Academy of Sciences* **320**, 28-42 (1979).
21. Buser, H. R. Polychlorinated dibenzo-*p*-dioxins and dibenzofurans: formation, occurrence and analysis of environmentally hazardous compounds, thesis, Department of Organic Chemistry, University of Umea, Sweden (1978).

Index

1,2,3,4,7,8-hexachlorodibenzo-*p*-dioxin, 31
1,2,3,6,7,8-hexachlorodibenzo-*p*-dioxin, 52
1,2,3,7,8-pentachlorodibenzo-*p*-dioxin, 5, 15, 31
1,2,3,7,8,9-hexachlorodibenzo-*p*-dioxin, 35, 52, 83
1,2,4,5-tetrachlorobenzene, 6
2,3,4,6-tetrachlorophenol, 6, 11
2,3,7-trichlorodibenzo-*p*-dioxin, 5, 30
2,3,7,8-TCDD: *see* dioxin
2,3,7,8-tetrachlorodibenzo-*p*-dioxin: *see* dioxin
2,3,7,8-tetrachlorodibenzofuran, 8, 9, 14, 31
2,4-D, 27, 40, 78, 129, 148, 149, 150, 151, 176, 188
2,4-dichlorophenol, 6
2,4-dichlorophenoxyacetic acid: *see* 2,4-D
2,4,5-sodium trichlorophenate, 7, 8, 9
2,4,5-T, 27, 128, 137, 237, 240
  ban, 41, 78, 80, 132, 179
  birth defect link
    40ff., 159, 175–7
    animals, 36–40, 147, 158, 159, 163, 164
  cancer link, 52–3, 102, 133, 175–6
    lymphatic system, 178
    soft tissue sarcoma, 52, 101, 178, 179
  chemistry, 5ff.
  chromosome damage link, 168
  Environmental Protection Agency (US)
    hearings, 173–4, 241
    study, 174–5
  incineration, 26
  kidney damage, animals, 37, 158

2,4,5-T (*cont.*)
  manufacture, 1, 9, 46, 81, 98, 110, 119, 121–2, 126, 132, 230
  military use, 149–50, 151, 187ff.
  perinatal mortality link, 41
  restrictions on use, 147, 159
  Vietnam, 147, 159, 162
  spontaneous abortion link, 41, 132, 174–5
  stillbirth link, 41
  tonnage used, UK, 177
2,4,5-trichlorophenol: *see* trichlorophenol
2,4,5-trichlorophenoxyacetic acid: *see* 2,4,5-T
2,4,5-trichlorophenoxyethanol, cancer link, animals, 51
2,4,6-trichlorophenol, 6, 11
2,7-dichlorodibenzo-*p*-dioxin, 28, 52
2,8-dichlorodibenzo-*p*-dioxin, 5, 30, 42
3-indole acetic acid, 148
3-methylcholanthrene, 55
4 androstene-3, 17-dione, 55
4 pregnene-3, 20-dione, 55
5α-androstane-3α, 17β-diol, 55
7,8-dihydroxy-9,10-epoxy-7,8,9,10-tetrahydrobenzo(*a*)pyrene, 51
7-ethoxycoumarin *o*-deethylase, 54, 56
12th Special Operations Squadron, U.S. Air Force, 147
AAAS: *see* American Association for the Advancement of Science
ABC Committee (US), 148
abnormal hair growth, 55, 129, 137, 138
abortion issue, Seveso, 212, 214–216
abortion law, Italy, 214
Abrams, General Creighton, 162

acetic acid, 78
acne, 129
  hexachlorophene treatment, 69
  see also chloracne
activated charcoal, 28
Adamo, Salvatore, 200
adenocarcinoma, animals, 51
adenoma, animals, 52
adipose tissue, animals, 29, 53
adrenals, 85
  animals, 29
  cancer, animals, 47
African Blood Lily, 46
Agent Blue, 151, 163
Agent Green, 151
Agent Orange, 1, 2, 8, 29, 127, 132, 137,
    151-2, 158, 164, 169, 170, 179,
    188, 200, 241
  birth defect link, 40, 158, 171, 172, 240
    animals, 169
  cancer link, 240
    colon, 170
    liver, 53, 168
    soft tissue sarcoma, 171-2
  child deaths, 40
  chromosome damage link, 46, 167-8
    animals, 169
  dioxin content, 164
  disposal
    chlorinolysis, 27
    incineration, 127-8, 162-3, 176
  eye damage, 211
  fertility problems, 169
    animals, 169
  heart disease link, 169
  libido loss, 169
  restriction on use, Vietnam, 158
  skin disease, 169
  spontaneous abortion link, 189
  stockpile
    Australia, 176-7
    US, 162-4
  suspension of use, Vietnam, 162
Agent Pink, 151
Agent Purple, 151
Agent White, 151, 188
AHH: *see* aryl hydrocarbon hydroxylase
Air America, 191
AKR mouse strain, 158
Albright & Wilson (UK), 78, 79
albumin, 35, 54
alcohol, bacteriacide, 73

alkaline phosphatase, 211
alopecia: *see* hair loss
"Alsea II" report, 174-5
"Amcide", 78, 79
American Association for the Advancement
    of Science, 153-7, 160, 162, 163,
    240
  Herbicide Assessment Commission, 159-
    62, 164, 167
  Pacific Division, 152
American Society of Plant Physiologists,
    152
Ames test, 42
ammonium sulfamate, 78
Amsterdam, 106, 108, 109
  Factory Inspectorate: 107
Amsterdam University, 18, 19
anemia, animals, 84
anencephaly, 176
anger, psychiatric problems, 36
angiosarcoma, animals, 51
aniline, 90
animal deaths, 123, 176, 200, 203
  see also individual named species
animal test:
  ED_{50} 37
  LD_{50}, 5, 9, 30, 31, 33, 69, 78
anorexia, 131
  animals, 32, 123
  see also appetite loss
anthracene compounds, 56
antibiotics, 111, 122, 209
antiseptics, 69ff.
appetite loss, 36; *see also* anorexia
Archbishop of Milan, 214
Aroclor 1260, 12
arsenic, 13, 151, 163
aryl hydrocarbon hydroxylase, 35, 54
Ashe, Dr. William, 90-1, 100
Association of Scientific Technical & Man-
    agerial Staffs (UK), 118, 121
ASTMS: *see* Association of Scientific,
    Technical & Managerial Staffs
    (UK)
ATPase activity, liver, 34
Australia, 40, 176-7
  Government, 172, 176, 177
  National Health & Medical Research
    Council, 19, 40, 176
  National University, 176
  Victoria Minister of Health, Consulta-
    tive Council, 176

Australia (*cont.*)
    Vietnam Veterans: *see* Vietnam veterans, Australia
auxin, 77, 148, 150
Axelrod, Dr. David, 230
azobenzene, 90
azoxybenzene, 90

B52 aircraft, 187, 189, 190, 191
B6D2F/J mouse strain, 57
bacteriacides, 6, 11, 69ff., 110, 126, 127, 198
bacteriological weapons: *see* biological weapons
Badische Anilin-und Soda-Fabrik: *see* BASF (W. Germany)
B(*a*)P: *see* benzo(*a*)pyrene
Barell, Emil, 208
BASF (W. Germany), 102–5, 107, 201
    mortality studies, workforce, 53, 104–5
Basle (Switzerland), 199, 201, 209
Baton Rouge (US), 126
Baucus, Senator Max, 191
Baughman, Robert, 167
Bayer, (W. Germany), 138
beef fat, dioxin levels, 29
benzaldehyde, 230
benzene, 230
    2,4,5-T structure, 78
    dioxin structure, 5, 25, 30, 90
benzo(*a*)pyrene, 51
    hydroxylase, 54
Bergamaschini, Dr. Ernesto, 199
BHK cell transformation assay, 42
Bikbulatova, Dr, 121–2
bile ducts, animals, 52
bilirubin, 211
Bilthoven (Netherlands), 12
Biogenics Corporation (US), 232
biological weapons, 148, 152, 153
Bionetics Research Laboratories (US), 148, 158–60, 163
biphenyls, 90
bird deaths, 123
birth defects
    2,4,5-T link, 40ff., 159, 175–7
    animals, 36–40, 147, 158, 159, 163, 164
    Agent Orange link, 40, 158, 171, 172, 240
    animals, 169
    attitude of Vietnamese, 161
    dioxin link, 1, 40–1, 130, 132, 133, 168–9, 206, 214, 215–8, 232, 239
    animals, 1, 36–40, 164, 214

birth defects (*cont.*)
    expected averages, 169, 216, 217
    herbicide spraying link, 159, 161, 162, 167, 168–9, 169, 171, 172, 189, 240
    hexachlorophene link, 74–5
    polychlorinated dibenzofurans link, 85
    see also individual named defects
Blackman, Professor G.E., 150
bladder damage, 125
    animals, 84
Blair, Ectyl, 173
Blank, Dr. Eric, 117
Bliss Waste Oil Company (US), 122, 126–7, 128
bloat, 220
blood fat levels, 36, 84, 85, 100, 116–8, 119, 122, 130, 136, 138–9
    animals, 34, 54
Blum, Barbara, 232
Boehringer (W. Germany), 89, 91–2, 112, 116, 131
Bohlem, Charles, 155–6
Bolsover (UK), 110, 112, 114, 139
bone defects, 86
Bookbinder, David, 114
Bordeaux mixture, 148
Bosshardt, Dr. Hans Paul, 16
Bowater Packaging Corporation (UK), 82
brain lesions, 71
Bratislava (Czechoslovakia), 19
"Breakfast" military mission, 191
breast cancer, animals, 47
breast milk, contaminated, 29, 86
British Colonial Office, 150
British Joint Technical Warfare Committee, 149
British Medical Journal, 73, 74
British Military Authorities, 149
bromine, 89
bronchial problems, animals, 52
bronchogenic lung carcinoma, 130–1
Brown, Michael, 231
brushwood killers, 78, 151, 178
Buckingham, William, 193–4
Bumb, Dr. Robert, 15, 18
Bunker, Ellsworth, 156, 157, 160, 162
burial, as dioxin disposal, 112, 114, 219, 222, 229
    at sea, 108
Burrows, Eric, 109, 139, 238
    inquest, 110
Buser, Dr. Hans Rudolph, 16

butanol, 78
Buxton, Professor P.A., 150

C-123 Provider aircraft, 147, 151, 158
C57BL/6J mouse strain, 37, 56, 57, 158
cacodylic acid, 151
Cambodia, 157, 187ff.
   alleged bombing, 3, 187ff.
   Government, 187, 188, 189
      Minister of Foreign Affairs, 187
   herbicide spraying, 3, 187ff.
   rubber plantations, 150, 187, 189
   2,4,5-T link, 52–3, 102, 133, 175–6
   2,4,5-trichlorophenoxyethanol, link, animals, 51
   adenocarcinoma, animals, 51
   adenoma, animals, 51
   adrenals, animals, 47
   Agent Orange link, 240
   angiosarcoma, animals, 51
   assays, short term tests, 41–6
   chromosome damage link, 212
   colon, 170
   digestive system, 53, 170
   dioxin link: see dioxin, cancer link
   ear duct, animals, 51
   fibrosarcoma, animals, 51
   genitourinary system, 53
   hematopoietic system, 53
   hard palate, animals, 47
   kidney, animals, 51
   leukemia, 84
      animals, 51
   liver, 53, 168
      animals, 47, 52
   lung, 53, 130–1
      animals, 47
   lymphatic system, 53, 178
      animals, 47, 51
   lymphocytic leukemia, animals, 51
   mammary gland, animals, 47
   pancreas, 108
      animals, 47
   pituitary, animals, 47
   register
      Seveso residents, 218
      Vietnam Veterans (US), 171
   skin, animals, 51
   smoking factor, 101, 134
   soft tissue carcoma, 52–3, 101, 133–4, 139, 171–2, 178, 179, 237, 240

Cambodia (cont.)
   testes, animals, 51
   thyroid, animals, 47
   trichlorophenol link, 52–3
   uterus, animals, 47
carbohydrate metabolism, 36, 122
carbon tetrachloride, 27, 28
carcinogens, 41, 117, 230
   latency period, 52, 53, 98, 131, 133, 168, 218
   see also cancer
cardiovascular disease: see heart disease
Carey, Governor Hugh, 231
Carreri, Dr. Vittorio, 201, 202, 204
Carter, President Jimmy, 170, 231, 233
cat deaths, 123, 125
cattle deaths, 84
caustic soda, 110
Cavallero, Dr. Aldo, 201–2, 204
CBS Broadcasting (U.S.), 170
CD-1 mouse strain, 37
Cefis, Professor Francesco, 215
cell culture tests, 47
Center for Disease Control (U.S.), 171
Central Intelligence Agency (U.S.), 189, 191
cerebral edema, 71
Cesano Moderno (Italy), 202, 218
charcoal, 28, 30
Charles-Rivers rat strain, 58
Charles University (Czechoslovakia), 129, 130
chemical weapons, 149–50, 152–3, 155
Chemie Linz (Austria), 138, 201
Chesterfield (UK), 110, 112
Chesterfield Royal Hospital (UK), 117
chick edema disease, 35, 83
chicken deaths, 12, 35, 83, 127
children
   chloracne, 104, 112, 125, 199, 239
   deaths, 40, 71ff., 127, 167
   dioxin poisoning, symptoms, 123–5
   herbicide exposure, symptoms, 167, 188
China, 156
chloracetic acid, 9
chloracne, 12–13, 33, 36, 84, 89ff., 98–100, 103, 108–9, 111, 116, 118, 121–2, 124, 131, 133, 136, 139, 164, 170, 171, 237
   abnormal hair growth, 129, 137, 138
   animals, 35, 91, 103
   causes, 89–90, 121

Index

chloracne (cont.)
  children, 104, 112, 125, 199, 239
  development, 85, 90-1, 137, 212
  duration, 89, 100, 104, 111, 112-3, 125, 129, 131, 135, 212-3
  heart disease link, 136
  odour, 90, 100, 137
  pigmentation, 122, 129
  susceptible complexions, 111
  treatment, 100, 111, 113, 122
chlorhexidene, 69ff.
  side effects, 69
chlorinated aliphatic hydrocarbons, 16
chlorinated aromatic hydrocarbons, 16
chlorinated benzene, 230
chlorinated compounds, 111, 130
chlorinated dibenzo-p-dioxins, 30, 84, 136
chlorinated dibenzofurans, 9, 31, 83
chlorinated diphenyls, 16, 121
chlorinated hydrocarbons, 198, 229
chlorinated naphthalenes, 103, 121
chlorinated-o-phenoxyphenols, 19
chlorinated phenols, 6, 9, 11, 18, 26, 81, 82, 131, 132, 136, 198
chlorinated toluene, 230
chlorine, 5, 13, 16, 18, 19, 30-1, 31, 52, 83, 89-90
chlorinolysis, 27
  dioxin disposal, 27-8
chloroform, 28
chlorophenol: see pentachlorophenol
cholesterol, 34, 36, 54, 84, 85, 100, 117, 118, 122, 130, 138
cholic acid, 30
chromatin fusion, 46
chromatography, 10, 17
chromosome damage, 232-3
  2,4,5-T link, 168
  Agent Orange link, 46, 167-8
  animals, 169
  cancer link, 212
  dioxin link, 46-7, 105, 117, 167-8, 211-12, 232
  animals, 46, 211-12
Church, Senator Frank, 191
C.I.A.: see Central Intelligence Agency (US)
Cimmino Commission, 220
Cimmino, Professor Aldo, 220
circulatory disorders, 102
Civel, Dominique, 72
civilian spraying programs (US), 147

cleft lip, 177
cleft palate, 37, 40, 162, 167, 169, 240
  animals, 37, 57
Clemm, August, 102
Clemm, Carl, 102
Cleveland Clinic Foundation (U.S.), 89
Clifton (U.S.), 198
"Clophen A 60", 12
club foot, 176
Clutterbruck, R., 150
Coalite & Chemical Products Limited (U.K.), 3, 34, 104, 109ff., 139, 201, 237, 238
  dioxin disposal, burial, 112, 114
  medical investigations, workers, 3, 111, 114, 116-7, 118-21, 139
colic, animals, 35
Colindale Public Health Laboratory (UK), 73, 75
Colombo, Cardinal, 214
colon, cancer, 170
Colorado State University (U.S.), 174
Commoner, Dr. Barry, 123
Como (Italy), 224
compensation claims, 113, 131, 165
  hexachlorophene poisoning, France, 72-3
  Love Canal residents, 223-4
  Seveso residents, 222
  Vietnam veterans (US), 132-3, 172-3, 179-80
congenital heart disease, 169
congenital malformations: see birth defects
conjunctivitis, 111, 129, 131
Constable, Professor John, 160
convulsions, 72
Cook, Robert, 160
cooking oil contamination, 12-14, 135
copper sulphate, 148
Cornell University (US), 167
coronary thrombosis: see heart disease
corticosteroids, 199
cosmetics industry, 11, 69
Cosmos Club (US), 154
Costle, Douglas, 173-4, 234
Crow, Dr. Kenneth, 90
Crummet, Dr. Warren, 19
Cutting, Dr. Robert, 161
cyclohexanone, 27
cystic kidney disease, 177
cytosolic receptor protein, 55-6
Czechoslovakia, 19, 129-31

δ-ALA-synthetase: *see* δ-aminolevulinic
    acid synthetase
d'Ambrosia, Dr. Francesco, 215
δ-aminolevulinic acid synthetase, 34, 54
Daily Telegraph (London), 113
Dalderup, Dr. L.M., 107
Dashiell, Dr. Thomas, 167
DBA/2J mouse strain, 56, 57
DBD: *see* dibenzo-*p*-dioxin
DCDD: *see* 2,7-dichlorodibenzo-*p*-dioxin
DDT, 81
de Carli, Dr. Luigi, 212
de Havilland Beavers, 150
deaths
    animals, 123, 176, 200, 203
        see also individual named species
    children and infants, 40, 71ff., 127, 167
    Yusho patients, 85
dechlorination, 19
defoliation: *see* herbicides, military use
dehydration, animals, 54
demolition, dioxin disposal, 107-8
Department of Health, Education &
        Welfare: *see* U.S. Government,
        Department of Health, Education
        & Welfare
Department of State: *see* U.S. Government,
        Department of State
depression, dioxin poisoning, 36
Derbyshire (UK), 115
Derbyshire County Council, Planning
        Committee (UK), 114-5
dermatitis, 121
Desio (Italy), 200, 202
di Domenico, Dr. Allessandro, 218
diabetes, 54
diarrhea, dioxin poisoning, 36, 188
Diamond Alkali (U.S.), 129
Diamond Shamrock (U.S.), 132, 172
dibenzofurans, 8, 9, 31, 84, 90
dibenzo-*p*-dioxin, 52
dibromomethane, 230
dicentric bridges, 46
dichlorodibenzo-*p*-dioxin, 5, 28, 30, 42, 52
dichlorophenol, 6
diethylnitrosamine, 58
digestive system, cancer, 53, 170
    dioxin poisoning, 36, 104, 131, 137, 202
dimethylbenzanthracene, 51
Dioguardi, Professor Nicola, 211
dioxin
    anticarcinogenic properties, 51-2

dioxin (*cont.*)
    assay, 167
    biochemical effects, 53ff.
    birth defects link, 1, 40-1, 130, 132, 133,
        168-9, 206, 214, 215-8, 232, 239
    birth defect link, animals, 1, 36-40, 164,
        214
    cancer link, 1, 41-2, 47-53, 57-8, 98,
        100-101, 105, 131, 136, 139
        adrenals, animals, 47
        animals, 1, 47-53, 84, 98, 105, 117, 134,
            136, 139, 175-6
        digestive system, 53, 170
        ear duct, animals, 51
        genitourinary system, 53
        hematopoietic system, 53
        kidney, animals, 51
        leukemia, 84
            animals, 51
        liver, 53, 168
            animals, 47, 52
        lung, 53, 130-1
            animals, 47
        lymphatic system, 53
            animals, 47, 51
        lymphocytic leukemia, animals, 51
        mammary gland, animals, 47
        pancreas, 108
            animals, 47
        pituitary, animals, 47
        skin, animals, 51
        smoking factor, 101, 134
        soft tissue sarcoma, 101, 133-4, 178,
            179, 237, 240
        testis, animals, 51
        thyroid, animals, 47
        uterus, animals, 51
    carcinogenicity: *see* dioxin, cancer link
    cell culture tests, 47
    chemistry, 5ff.
    chromosome damage link, 46-7, 105, 117,
        167-8, 211-2, 232
        animals, 46, 211-2
    cytology, 46
    cytosolic receptors, 55-6, 57
    food chain contamination, 167
    heart disease link, 100, 102, 108-9, 113,
        118, 120-1, 122, 130, 131, 133, 139
    low birth weight link, 231
    mutagenicity, 41-7, 58
    nuclear receptors, 56
    solubility, 25

Index

dioxin (*cont.*)
  spontaneous abortion link, 130, 217-8
    animals, 47, 123
  steroid receptors, 56
  tissue culture tests, 47
  toxicology, 25ff., 241
  see also 2,4,5-T, Agent Orange
dioxin decontamination measures, 103, 107, 111, 112
dioxin degradation: 25-8, 114, 218-9, 241
  microbial action, 25-6, 115
  soil, 25-6, 27, 115, 219
  sunlight, 27, 219
  ultraviolet light, 26-7, 115, 128, 221
dioxin disposal, 128, 218, 222
  burial, 112, 114, 219, 222, 229
  burial at sea, 108
  chlorinolysis, 27-8
  demolition, 107-8
  incineration, 26, 115, 127-8, 221-2
  oxidization, 28
  steam, 221
  solvents, 109, 221
  waste tips, 109, 115
dioxin excretion, 30
dioxin experiments, prisoners, 134ff.
dioxin exposure, compensation claims: *see* compensation claims
dioxin formation: incineration, 14-21, 26, 40, 163, 176, 241
dioxin levels, limits, 9-10, 164, 177, 179
dioxin poisoning, 36, 122, 124, 129, 131, 152ff.
  adipose tissue, animals, 29, 53
  anemia, animals, 84
  anorexia, 131
    animals, 32, 123
  appetite loss, 36
  bile duct, animals, 52
  birth defects: *see* dioxin, birth defect link
  bladder, 125
    animals, 84
  blood fat levels, 36, 100, 116-8, 119-20, 122, 130
    animals, 34, 54
  bronchial problems, animals, 52
  cancer: *see* dioxin cancer link
  carbohydrate metabolism, 36, 122
  chloracne: *see* chloracne
  chromosome damage: *see* dioxin, chromosome damage link
  colic, animals, 35

dioxin poisoning (*cont.*)
  conjunctivitis, animals, 35
  death, animals, 200, 203
  dehydration, animals, 54
  depression, 36
  diarrhea, 36, 188
  digestive system, 131, 137, 202
    animals, 36
  duodenal ulcer, 104
  edema, animals, 35
  endocrine system, 55
  energy loss, 36
  eyelids, animals, 35, 52
  fertility problems, animals, 47, 169
  fibrosis, 36
  hair growth, abnormal, 55, 138
  hair loss, 55
    animals, 58
  headaches, 36, 99
  hearing loss, 36
  heart disease: *see* dioxin, heart disease link
  hirsutism, 55, 138
  hormone metabolism, animals, 55
  immune system, 33, 36, 111, 117, 119, 120, 209, 211
    animals, 32-4, 57
  impotence, 100, 122
  insomnia, 36, 99, 100
  joint pain, 36
  joint stiffness, animals, 35-6
  kidney, 111, 135, 200
    animals, 35
  libido loss, 55, 100, 131
  liver, 36, 53, 103, 111, 116-7, 119-20, 122, 129-30, 131, 135
    animals, 11, 28, 34-5, 53, 54, 106-7
  low birth weight: *see* dioxin, low birth weight link
  lungs, animals, 32
  menstrual problems, animals, 47
  muscle pain, 36
  nail abnormalities, animals, 35
  nausea, 36, 99, 131
  nervous system, 36, 99, 100, 104, 122, 130, 131, 210
  pancreas, 36, 104, 108
    animals, 52
  placenta, animals, 54
  polyneuropathies, 36
  protein metabolism, 122, 130

dioxin poisoning (cont.)
  psychiatric problems, 36
    anger, 36
  reproductive system, animals, 35, 40, 47, 54, 107
  respiratory system, 36, 52, 100, 122
    animals, 32, 52
  salivary glands, animals, 52
  sensory disturbance, 36, 104
  spontaneous abortion: see dioxin, spontaneous abortion link
  stomach, 35, 104
  sugar metabolism, 130
  testes, animals, 35, 107
  thymus, animals, 32, 57, 107
  tonsillitis, 122
  urinary tract, 36
  uroporphyrinuria, 36, 129, 130, 200
    animals, 34
  vertigo, 99
  vomiting, 36, 99, 188
  weight loss, 36, 125
    animals, 32, 35, 84
diphtheria, 171
disablement pensions, 131
DMBA, 51
DNA, 41, 46, 51-2, 55, 56, 57
dog deaths, 123
Dow Chemical Company (US), 106, 107, 132-6, 201, 241, 242
  2,4,5-T hearings, 163-5, 175, 179, 241-2
  compensation claims, 132-3, 170-3, 241-2
  incineration report, 15-21
  medical investigations, workers, 46, 47, 51
  mortality studies, workforce, 101, 133-4, 178
  pregnancy study, employees' wives, 41
Down's syndrome, 169, 215, 217
Dubendorf (Switzerland), 198, 199, 201, 203, 204
Dubos, Dr. Rene, 154
DuBridge, Dr. Lee, 147, 159
duodenal ulcer, dioxin poisoning, 104
DuPont, 79
Dusseldorf University (W. Germany), 105
dust supression, 122-3, 126-7
dye production, 102

E. coli, 42
ear duct, cancer, animals, 51

East Africa, 150
ED_{50}, animal test, 37
edema, animals, 12, 35, 83
Edsall, Dr. John, 153
effluent discharge, 82
Egypt, 157
Ellipticine, 46
EMAS: see Health & Safety Executive (U.K.), Employment Medical Advisory Service
Employment Medical Advisory Service: see Health & Safety Executive (U.K.), Employment Medical Advisory Service
endocrine system, 55
Environmental Protection Agency (U.S.), 9, 71, 79, 84, 127-8, 132, 163, 164, 176, 230, 232-4, 240
  2,4,5-T hearings, 135, 173-4, 179, 241
  2,4,5-T study, 41, 174-5
  Scientific Sub-Committee, 175
EPA: see Environmental Protection Agency (U.S.)
epithelium, 52
Escherichia coli, 42
esters, 78
ethanol, 28, 78
ethylene glycol, 7, 102, 110, 222
explosives, forest clearance military use, 149
eye damage, 211; see also conjunctivitis

Faccio, Adele, 215
Factory Inspectorate (UK), 110, 238
Fara, Professor Gaetano, 211, 216, 224
Federal Common Law (US), 173
Federation of American Scientists, 152, 159
fertility problems, 169
  animals, 47, 169
fibroblasts, 47
fibrosarcoma, animals, 51
fibrosis, 36
fibrous tissue sarcoma, 134
Field, Barbara, 177
Find, Judge Edward B., 173
Finland
  forestry workers, 52, 240
  National Board of Health, 178
fish, dioxin susceptibility, 36, 77
fish contamination, dioxin, 167
Fishook (Cambodia), 187, 189, 190, 191, 193

Index

Flahault, Hubert, 72
flavourings, manufacture, 208
Florida, 25, 29, 115, 149, 150
Food & Drug Administration (U.S.): *see* US. Government, Food & Drug Administration
food chain contamination, dioxin, 167
forest areas, Vietnam, 147, 151, 156
forestry, 78, 80
Forestry Commission (UK), 137
   Research Department, 79, 132
Forestry workers
   Finland, 52, 240
   Sweden, 52-3, 177-9, 240
   U.K., 137
formaldehyde, 11
Fort Detrick (U.S.), 148, 149, 150
Foster, John S., 154, 155
frameshift mutagens, 42
France, 71-3, 104, 127, 131, 148, 189
Frentzel-Beyme, Dr. R., 104
Freedom of Information Act (US), 189
Fukoaka (Japan), 12
fungicides, 6

Galston, Professor Arthur, 152-3, 159
gamma glutamyl transpeptidase, 117, 211
gas liquid chromatography, 10, 106
gastrointestinal tract, cancer, 53
genetic damage: *see* chromosome damage
Geneva: 71, 208, 224
Geneva Protocol (1925), 153
genitourinary system, cancer, 53
Genoa (Italy), 215
George Washington University (US), 174
Ghetti, Dr. Giuseppi, 201, 203, 224
Giorno, 201
Giovanardi, Professor Augusto, 220, 224
Givaudan (Switzerland), 8, 11, 71-2, 110, 197ff., 237-8
Givaudan, Leon, 208
Givaudan, Xavier, 208
Globe (U.S.), 176
Glossina sp, 150
Gluck, Louis, 74
glucocorticoid hormones, 56
glucuronyl transferase, 54
glutamic oxaloacetic transaminase, 34, 36, 210
glutamic pyruvic transaminase, 34, 36
"Glyphosate", 78, 79, 80
Golfari, Cesare, 202

Gothenberg (Sweden), 74
Gravesend (U.K.), 82
grey seal, 14
Gropp, Professor Alfred, 215
growth retardation, 86
Guardian (London), 113
Gulfport (U.S.), 128
guppies, 36

Haemanthus Katherinae, 46
hematopoietic system, cancer, 53
hair growth, abnormal, 36, 55, 129, 137, 138
hair loss, 55
   animals, 35, 55
Hall, Dr. Peter, 176-7
halogenated hydrocarbon, 57, 210
halogens, 89-90
Hamburg (W. Germany), 89, 91
Hampshire College (U.S.), 152, 188
Hanoi (Vietnam), 168
Hardell, Dr. Lennart, 178
hard palate cancer, animals, 47
hardwood forests, Vietnam, 161
Harpenden (U.K.), 137
Hart, Senator Philip, 126
Harvard University (U.S.), 153, 158, 160
Harvey, Judge James, 175
Hawaii (US), 150, 189
Hayward, Thomson: *see* Thomson Hayward (US)
HDL: *see* high density lipoprotein
headaches, dioxin poisoning, 36, 99
Health and Safety at Work Act (UK), 1974, 119-120
Health and Safety Executive (UK), 84, 114, 116-9, 121, 136, 137, 139, 238
   Employment Medical Advisory Service, 119-20, 136
heart disease
   Agent Orange link, 169
   congenital, 169
   chloracne link, 136
   dioxin link, 36, 100, 102, 108-9, 113, 118, 120-1, 122, 130, 131, 133, 139
   see also blood fat levels
hemangioma, 217
hepatic enzymes, 36, 54
hepatic porphyria, 34
hepatoma, 47
hepatocellular adenoma, animals, 52
hepatocellular carcinoma, animals, 47, 52

hepatocytes, 34
heptachlorodibenzo-*p*-dioxin, 19, 82, 136
Herbicide Assessment Commission: *see* American Association for the Advancement of Science: Herbicide Assessment Commission
herbicide spraying, birth defect link, 159, 161, 162, 167, 168-9, 171, 172, 189, 240
herbicides, military use, 2, 9, 148-50
  Cambodia, 3, 187ff.
  Vietnam, 132, 147ff.
  see also individually named herbicides
Hercules Incorporated (US), 132, 172
Herxheimer, Dr. Karl, 89
Hevea brasiliensis, 150
hexachlorodibenzo-*p*-dioxin, 19, 31, 35, 52, 82, 83, 84, 136
hexachlorophene, 1, 11, 69ff., 81, 110, 126, 127, 198
  acne treatment, 69
  birth defects unit, 74-5
  impetigo treatment, 69
  infant deaths, 71-3, 74-5, 127
  restrictions on use, 75
  side-effects, 71-3
Hickey, Dr. Gerald, 167
high density lipoprotein, 54, 117, 136
high explosives, military use, forest clearance, 149
Higuchi, Kentaro, 85
Hinkle, Charles W., 191
histidine, 42
histiocytoma, animals, 51
Ho, Professor Pham-Hoang, 166
Ho Chi Minh Trail, 193
Hoffman, Fritz, 208
Hoffmann-LaRoche (Switzerland), 11, 110, 197, 199, 203-6, 208-9, 219, 221, 222, 224, 225
Hoffman Taft Inc. (US), 126
Holland, 115, 139
Holmesburg Prison (US), 134-6
Holmsted, Professor Bo., 92
Hooker Chemical Co. (US), 138, 229, 231, 233, 234
hormone metabolism, 55
horses
  deaths, 123
  dioxin poisoning, 35, 123
horse arenas, contamination, 123-8

HSE: *see* Health & Safety Executive (UK)
Hungary, 177
Hurley, Jonathan, 100
Hutzinger, Dr. Otto, 18
Hyde Park (US), 234
hydrochloric acid, 102
hydropericardium heart sac, 35
hyperbilirubinemia, 34
hypercholesterolemia, 34, 36, 54
hyperlipidemia, 54
hyperpigmentation, 36
hypertriglyceridemia, 36
hypochlorite, 28
hyperproteinemia, 34
hypospadia, 217

IARC: *see* International Agency for Research on Cancer
ICMESA (Italy), 197-8, 199, 201, 204, 207, 209, 219, 221, 222, 223, 224, 225
ICMESA, workers strike, 200
Idaho (US), 191
immune function tests, 210
immune system, 33, 36, 111, 117, 119, 120, 209, 211
  animals, 32-4, 57
immunoglobulins, 54
immunostimulants, 209
impetigo, hexachlorophene treatment, 69
impotence, 100, 122
incineration, 229, 241
incineration
  Agent Orange, 127-8, 162-3, 176
  dioxin disposal, 26, 115, 127-8, 221-2
  dioxin formation, 14-21, 26, 40, 163, 176, 241
incinerator ship: "Vulcanus," 127-8, 163
Independent Petrochemical Company Inc. (US), 126-7
industrial hygiene: *see* occupational hygiene
Industrie Chimiche Meda Societa Anonima: *see* ICMESA
infant deaths, 40, 71ff., 127, 167
  see also perinatal mortality
Inouye, Senator Daniel K., 189, 191
insomnia, 36, 99, 100
International Agency for Research on Cancer, 8-9, 104, 105
iodine, 89-90
iron, 54
ischemic heart disease, 136, 139

Index

Instituti Clinici di Perfezionamento (Italy), 215
Instituto Superiore di Sanita (Italy), 218
Italy
  Government, 225
  Judiciary, 222, 225
  Ministry of Health, 202, 219-20
  Ministry of Health, Cimmino Commission, 220
  Parliament, 215
  Parliamentary Commission of Enquiry, Seveso, 207, 208, 210
  Radical Party, 215
  terrorism, 224

Japan, 79, 149, 79, 85
  contaminated cooking oil, 12-14, 135
Jirasek, Professor L., 130
Johnson, Anita, 158-9
Johnson, President Lyndon B., 152-3
Johnston Island (Pacific), 162, 163, 164
joint pain, 36
joint stiffness, animals, 35-6
Jones, Gary, 179
jungle rot, 171

Kampong Cham (Cambodia), 187, 188, 189
Kampuchea: *see* Cambodia
"Kanechlor 400", 13
Kanegafuchi Chemical Industry Co. (Japan), 13
Karolinska Institute (Sweden), 92
Kennedy, President John F., 151
Kenya, 150
keratin, 90
Kerr, Charles, 177
Khymer-Vietnamese frontier, 187
kidney cancer, animals, 51
kidney damage, 111, 135, 200
  animals, 35, 37, 158
kidney disease, 177
kidney function tests, 200
Kimbrough, Dr. Renate, 123-4, 126, 127
Kissinger, Dr. Henry, 191, 193
Kligman, Dr. Albert, 134-6
Klingberg, Professor Marcus, 217
Korean War, 149
Kraus, Dr. E.J., 148
Krek (Cambodia), 187
"Krenite", 78, 79

Kuala Lumpur, 150
Kyushu University (Japan), 12

La Porte, Dr. Vito, 218, 219
La Berge, Walter, 191, 193
Lamb, Dr. James, 169
Lancet, 118
landfills, 115, 125, 128, 229ff.
Laos, 3, 193-4
Lapland, 176
lard oil, 25
Larsson, Dr. Sune, 178
latency period, carcinogens, 52, 53, 98, 131, 133, 168, 218
Lavorel, Dr. Jean, 188
LD_{50}, animal test, 5, 9, 30, 31, 69, 78
Lee, Dr. Donald, 137-8
Lemos, Rear Admiral William, 159
leukemia, 84
  animals, 51
Lexington Blue Grass (US) army depot, 84
libido loss, 55, 100, 131, 169
lignin, 18
limb paralysis, animals, 71
lipase, 104
lipid metabolism: *see* blood fat levels
liposarcoma, 101
Litton Industries (US): Bionetics Research Laboratories, 148, 158-60, 163
liver, ATPase activity, 34
liver cancer
  Agent Orange link, 53, 168
  dioxin link, 168
  animals, 47, 52
liver cirrhosis, 130
liver damage, 34, 36, 53, 103, 111, 116-7, 119-20, 122, 129-30, 131, 135, 211
  animals, 11, 28, 34-5, 53, 54, 106-7
liver function tests, 200, 210, 211
liver microsomal enzymes, 41
Lombardy Region (Italy)
  Government, 201, 202, 203-4, 214, 215, 217, 220, 225
  Health Assessors Council, 201
  Health Authority, 199, 206, 208, 210, 214, 215, 220
  Health & Epidemiology Commission for Seveso, 210
  Special Office for Seveso, 213, 217, 218, 219, 224
Longhorne, Lieutenant Colonel Jack, 147

lotio alba, 100
Love Canal (US), 3, 229ff.
Love Canal
  chemical waste, 229
Love Canal
  residents
    compensation claims, 223-4
    evacuation, 231, 233
    medical investigations, 231-3
    offspring
      birth defects, 232
      low birth weight, 231
Love, William, 229
low birth weight, 86, 231
Lübeck Medical School (W. Germany), 215
Ludwigshaven (W. Germany), 53, 105, 107
Lugarno (Switzerland), 204
lung cancer, 53, 130-1
  animals, 47
lung damage, animals, 32
lymphatic system cancer, 53, 178
  animals, 51
lymphocytes, 32, 33, 117, 211
  transformation, 33
lymphocytic leukemia, animals, 51
lymphoma, 47, 178
Lyons (France), 104

McClanahan, Ivan, 99
McConnell, Dr. Eugene, 84
McGraw-Hill, 157
MacNamara, Robert, 154, 193-4
MAFF: see Ministry of Agriculture Fisheries & Food (UK)
Malaya, 149-50
malformations: see birth defects
Malgrati, Fabricio, 198, 200
malignant fibrous histiocytoma, 101, 134
malignant histiocytoma, animals, 51
malignant lymphoma, 178
mammary cancer, animals, 47
Mangiagalli Hospital (Italy), 215
mangrove forests, Vietnam, 147, 156, 161, 166
Mantel, Professor Nathan, 174
Mariano Comense Hospital (Italy), 199
Mario Negri Institut (Italy), 220
Marshall, R.P., 116
Martin, Dr. Jenny, 117-8, 119, 120
Martindale Pharmacopoeia, 71
mass spectrometry, 10-11

Matarasso, Leon, 188
maternity clinic, bacteriacide use, 71, 73
May, Dr. George, 110-1, 114, 115-6, 120
Mead, Dr. Margaret, 157
Meda (Italy), 198, 200, 201, 204, 209, 224
medical investigations
  forestry workers, Sweden, 52-3, 177-9, 240
  Lexington residents (US), 84
  Love Canal residents, 231-3
  Seveso residents, 2, 209-10
    offspring, 215-8
  Vietnam veterans
    (Australia), 177
    (US), 171
  workers, 33, 52-3, 89, 104-5, 121-2, 129
    Coalite & Chemical Products Ltd. (UK), 3, 111, 114, 116-7, 118-21, 139
    Monsanto Chemical Company, 84, 90-1, 101-2, 136-7
    Philips-Duphar, 108-9
    Yusho patients, 12-13
Mekong Delta (Vietnam), 147
menstrual problems, 85
  animals, 47
'Menu' military mission, 191
mercury-based compounds, 82
Meselson, Professor Matthew, 153, 158-9, 160-1, 162, 163, 166-7
methanol, 26, 28, 102
methanolic sodium hydroxide, 6-7
methylene bis (thiocyanate), 82
Michigan (US), Department of Natural Resources, 15
microagroeco-system chamber, 27
microbial action, dioxin degradation, 25-6, 115
microcephaly, 217
microsomal enzymes, 41
microsomal monoxygenase activity, 57
Midwest Research Institute (US), 154-5
Milan (Italy), 198, 199, 201, 202, 204, 206, 220, 221, 224
  Archbishop, 214
  Dermatology Clinic, 212
  Institute of Hygiene, 220
  Instituto Clinici de Perfezionamento, 215
  Mangiagalli Hospital, 215
  Mario Negri Institut, 220

Milan (Italy) (*cont.*)
  Niguarda Hospital, 199, 200, 202
  Provincial Laboratory of Hygiene & Prophylaxis, 201
Milan University, 202, 204, 211
Military Assistance Command for Vietnam: *see* United States, Government, Department of Defense, Military Assistance Command for Vietnam
Military Authorization Bill, 1970 (US), 163
Military Authorities (UK), 149
military herbicide use, 1, 2, 3, 8, 40, 127, 132, 137, 147ff., 187ff., 200, 211, 240
Mimot (Cambodia), 187
Minarick, Dr. Charles, 149
Ministry of Agriculture, Fisheries & Food (UK), 77, 137–8, 177
  Pesticides Advisory Committee, 77, 177–8
Ministry of Technology (UK), 110
miscarriage: *see* spontaneous abortion
Missouri (US)
  Health Department, 127
  horse arenas, contamination, 122–6
Mitchell, John. 148
mitosis, 46
Monsanto Chemical Company (US), 79, 98–102, 105, 116, 132, 139, 172, 237
  Newport, South Wales (UK), 136–7
  medical investigation, workers, 84, 90–1, 136–7
  mortality studies, workers, 53, 102, 134, 178
Montagnard people (Vietnam), 40, 161, 167
Montana (US), 188
Moore, Professor John, 9
Morhange (France), 71–3
mortality studies, workers, 240
  BASF (W. Germany), 53, 104–5
  Dow Chemical Company (US), 133–4, 178
  Monsanto Chemical Company (US), 53, 102, 134, 178
  Philips-Duphar (Netherlands), 108–9
Morton (UK), 115
multinucleate cells, 34, 46
Munn, Dr. Alex, 137
muscle pain, dioxin poisoning, 36
Mussolini, Benito, 214

mutagenicity, dioxin, 41–7, 58
mutagens, 41–7
myocardial infarction: *see* heart disease

"Nader's Raiders", 158
naphthalene, 90
nail abnormalities, animals, 35
National Academy of Sciences (US), 147, 148, 154, 163, 165, 167
  ABC Committee, 148
National Cancer Institute (US), 158
National Disarmament Conference (18th) Geneva, 155
National Forest Products Association (US), 174
National Institute of Environmental Health Sciences (US), 9, 84
National Institute of Public Health (Netherlands), 17
National Institute for Occupational Safety & Health (US), 84, 138
National Law Journal, 172
National Research Council (US), 148, 154
National Union of Agricultural & Allied Workers (UK), 177, 178
Nature, 119, 120, 139
nausea
  dioxin poisoning, 36, 99, 131
  polychlorinated dibenzofurans, 85
n-butyl esters, 149, 151
Needham, Charles, 113–6, 119
NEPACCO: *see* North Eastern Pharmaceutical & Chemical Company Inc. (US)
nervous system
  damage, hexachlorophene, 72
  dioxin poisoning, 36, 99, 100, 104, 122, 130, 131, 210
Netherlands, 12, 19, 106–9, 163, 201
neural tube defects, 40, 176, 177
  see also spina bifida
New South Wales (Australia), 177
New York (U.S.), 230, 231, 233
  Department of Environmental Conservation, 231
New York Times, The, 160
New Zealand, 40
  Department of Health, 40, 176
Newark (US), 129
Newport, South Wales (UK), 136
Niagara Falls (US), 138, 229ff.

Niagra Gazette, 231
nickel, 16
Niguardia Hospital (Italy), 199, 200, 202
Nissan (Japan), 79
nitrogen, 54
nitromethane, 28
Nixon, President Richard M., 147, 159, 165, 191
NMRI mouse strain, 37
Nobel Laureates, herbicide petition, 153
Noe, Luigi, 214, 219, 220, 222, 224–5
North Island (New Zealand), 40
North Eastern Pharmaceutical & Chemical Company Inc. (US), 126–7, 128
North Vietnam, 156, 167, 168
Norway, 79–80
nuclear chromatin, 56
nuclear receptors, dioxin, 56
nurseries, hexachlorophene use, 71

Occidental Petroleum Corporation (US), 233
occupational hygiene, 9, 19, 95, 103, 111, 130, 131–2, 138, 239, 242
Ocean Combustion Services (Netherlands), 163
octachlorodibenzo-*p*-dioxin, 19, 21, 42, 82, 136
Office of the Government Chemist (U.K.), 10
oil-based herbicide, 78, 151
olive oil, 27
"Operation Ranch Hand", military mission, 147, 151, 171, 172
Oregon (US), 41, 132, 174–5
Oregon State University (US), 175
organic solvents, 109
organobromine, 82
organosulphur, 82
Orians, Dr. Gordon, 157–8, 163
orthochlorobenzene, 110
orthodichlorobenzene, 25
Osservatore Romano (Vatican City), 214
oxidization, dioxin disposal, 28

PAC: *see* Ministry of Agriculture, Fisheries & Food (UK), Pesticides Advisory Comittee
Pacific Islands, 149
Paecilia reticulatus, 36
Palme, Olaf, 165

pancreas
 cancer, 108
 animals, 47
 dioxin poisoning, 36, 104, 108
Paoletti, Paolo, 198, 200, 224
papermaking industry, 81
parenchymal cells, 34
Pavia (Italy), 212
PCB: *see* polychlorinated biphenyl
PCDD: *see* polychlorinated dibenzo-*p*-dioxin
PCDF: *see* polychlorinated dibenzofurans
PCP: *see* pentachlorophenol
pentachlorodibenzo-*p*-dioxin, 5, 15, 19, 31
pentachlorofurans, 13
pentachlorophenate, 136
pentachlorophenol, 6, 11, 19, 36, 81, 82–4, 136, 178
 chicken deaths, 35, 83
 cattle deaths, 84
 wood preservative, 84, 91
Percy, Senator Charles, 172
perfume manufacture, 208
perinatal mortality, 41
peripheral neuropathy, 100
peritoneal cavity, 35
Pfeiffer, Dr. Egbert W., 152, 153, 157, 158, 162, 163, 165, 188, 191
phenobarbital compounds, 56
"Phenoclor DP 60", 12
phenol, 90, 91, 110
phenolic hydroxyl, 30
phenoxy acids, 78, 79, 80, 178
phenoxy herbicides, 148, 149
phenoxyphenols, 19
Philips-Duphar (Netherlands), 106, 201
 dioxin decontamination, 107–9
 medical investigations, workers, 108–9
 mortality studies, workers, 108–9
Phnom Penh (Cambodia), 187
photochemical decomposition: *see* ultraviolet light, dioxin degradation
photodegradation: *see* ultraviolet light, dioxin degradation
photolysis, 21
photosensitivity, 71
phytohemaglutin, 33
Picciano, Dr. Dante, 232
picloram, 151, 188
pigmented skin, 36, 86, 122, 129
pineapple crops, 82

Index

Pioneer aircraft, 150
pituitary cancer, animals, 47
plancenta, dioxin poisoning, animals, 54
Plain of Reeds (Vietnam), 157
plant hormones, 77, 148, 150
Plant Pathology Laboratory (U.K.), 137
polychlorinated biphenyls, 12, 13, 14, 15, 81, 85
   poisoning symptoms, 85-6
polychlorinated dibenzo-*p*-dioxin, 5, 6, 19, 21, 52, 55, 89, 95, 241
polychlorinated dibenzofurans, 8, 9, 11, 12, 13, 14-15, 18, 85-6
   birth defects link, 85
   poisoning, symptoms, 85-6
polyneuropathies, 36
polythylene glycol, 28
polyvinylchloride, 5
porphyria, 34
porphyrins, 36, 200
Port of London Authority (UK), 82
potassium trichlorophenate, 137
Prague (Czechoslovakia), 129
Pran Buri (Thailand), 150
Pratt, Judge George, 173
predioxins, 19
Price, Don, 154
primary liver cells, 47
Printing Industries Research Association (UK), 82
propanol, 78
protein metabolism, disordered, 122, 130
protein synthesis, 55
proteolytic enzymes, 56
Provincial Laboratory of Hygiene & Prophylaxis (Italy), 201
psychiatric problems, dioxin poisoning, 36, 91, 130, 137
Puccinelli, Professor Vittorio, 202, 212
Puerto Rico, 150
Purdue University (US), 223
PVC: *see* polyvinylchloride
pyrene compounds, 56

Queensland (Australia), 40, 176

Radical Party (Italy), 215
railroad workers, Sweden, 52
rainbow trout, 36
"Ranch Hand" military mission, 147, 151, 171, 172

Rappe, Professor Christopher, 5, 8, 9, 11, 14, 16-18
reactor accidents, 36, 95ff., 197ff., 238, 282-4
Reagan, President Ronald, 179
Reggiani, Dr. Giuseppi, 199-201, 203-4
Rehder, Dr. Helga, 215
reindeer deaths, 176
renal proximal tubule cells, 54
reproductive system, 85-6
   animals, 35, 40, 54, 170
respiratory system, 36, 85, 100, 122
   animals, 32, 52
Reutershan, Paul, 170
Rezzanico, Ugo, 209
Rhone-Poulenc Pont de Claix (France), 131-2
rice oil contamination, 12-14, 135
Richards, Professor Paul, 166
Rivolta, Dr. Vittorio, 201, 202-4, 206, 220, 224
RNA, 55
Roberts, Dr., 199
Rocca, Francesco, 198, 200, 202
Roche: *see* Hoffmann-LaRoche
Rohrborn, Professor G., 105
Rollins Environmental Service Company (US), 126
Rollins Purle Inc. (US), 126-7
Rome (Italy), 211, 218, 219-20
Rossyn, General, 160
Rostow, Walt, 151
Rowe, Dr. V.K., 134-6
rubber plantations, 150, 187, 189
Rupp, Rudolph, 224
Rusk, Dean, 193-4
ruthenium tetroxide, 28

S-9 mix, 41-2
Saigon (Vietnam), 156, 157, 160, 161, 162
   Children's Hospital, 162
   U.S. Embassy, 156, 162
Saigon University, 166
salivary gland, animals, 52
Salmonella, 32-3, 42, 58
Sambath, Huot, 187
Sambath, Dr. Jorg, 7, 198-9, 204, 223, 224
Sanchioni, Dr. Laura, 215
"Santobrite," 136
Savage, Dr. Eldon, 174

Scandinavia, 11; *see also* individual named countries
Schistosomiasis, 82
Schultz, Dr. Karl, 89, 91–2, 111–2, 116
Science, 18
Science Advisory Committee (US), 164
Scientific Research, 157, 158
Scientist & Citizen, 153
Selinger, Dr. Ben, 176–7
Sencar mice strain, 51
sensory disturbance, dioxin poisoning, 36, 104
Setico (France), 71–3
Seveso (Italy), 1, 3, 9, 25, 27, 95, 105, 107, 109, 110, 113, 116, 122, 124, 128, 139, 197ff., 229, 230, 237, 238, 242
Seveso
  abortion issue, 212, 214–5, 239
  agriculture, 219, 220, 222
  Decontamination Commission, 220–1, 224
  Health & Epidemiology Commission, 216, 224
  ICMESA workers, dioxin exposure, 200
  local doctors, 214, 239
  local health officer, 197, 198, 199, 224
  map, contaminated areas, 201, 203–5
  residents
    cancer register, 218
    children, 202, 207, 210–11
    chloracne, 212, 214
      skin disease, 199, 206
    compensation claims, 222
    evacuation, 2, 201, 203, 204, 206, 207, 208, 209
    medical investigations, 2, 209–10
    offspring, medical investigation, 215–8
    pregnant women, 206, 214–8, 225
    spontaneous abortion, 215, 217–8
  zoning system, 204–7
    Zone A, 124, 204, 206, 207, 210, 211, 218, 219, 220–1, 222, 229
    Zone B, 205, 206, 210, 211, 219, 220, 221
    Zone R, 205, 219, 220
SGOT: *see* glutamic oxaloacetic transaminase
SGPT: *see* glutamic pyruvic transaminase
Shawcross, William, 189, 190, 191
sheep deaths, 220
Sheffield (UK), 113
Sheffield University (UK), 117
shellfish contamination, dioxin, 167
short term tests, cancer assays, 41–6
Sihanouk, Prince, 187
"Silvex", 26
silviculture: *see* forestry
Site Engineering Services (UK), 112
skin cancer, animals, 51
skin disease, 85, 121, 169
  chloracne: *see* chloracne
  diphtheria, 171
skin
  fragile, 129
  sensitive, 69, 71
Skinner, Dennis, 116
sleeping sickness, 150
sleep disturbance: *see* insomnia
slimicides, 6, 81
smokeless fuels, 109
smoking factor, dioxin, cancer link, 101, 134
snail control, 82
Society for Social Responsibility in Science (US), 157
Södertälje Hospital (Sweden), 74
sodium-2-hydroxyethoxide, 7
sodium chloride, 90
sodium hydroxide, 7, 107
sodium trichlorophenate, 7, 8, 9
soft tissue sarcoma, 52, 101, 133–4, 139, 171–2, 178, 179, 237, 240
soil, dioxin degradation, 25–6, 27, 115, 219
solid smokeless fuels, 109
solvents, dioxin disposal, 109, 221
somatic cells, 212
Sorge, George, 91–2
South America, 163
South American leaf blight, 150
South Vietnam, 147, 148, 151, 155–6, 158, 160–2, 165, 167, 168
  Central Highlands, 161
  Ministry of Health, 161
Soviet Union: *see* USSR
Spallino, Antonio, 224
spectrophotometry, 106
Spencer, Dr. David, 126
Spillane, Anthony, 112
spina bifida, 40, 162, 167, 169, 176, 177, 240
Spolana (Czechoslovakia), 129–31

Index

spontaneous abortion, 169
  2,4,5-T link, 41, 132, 174-5
  Agent Orange link, 189
  dioxin link, 130, 217-8
  animals, 47, 123
spotted sunfish, 29
Sprague-Dawley rat strain, 55, 158
squamous cell carcinoma animals, 47
SSRS: *see* Society for Social Responsibility in Science (US)
Stamford (US), 208
staphylococci, 11, 69, 71, 73, 74, 75
Star (Sheffield), 113-4, 120
Statens Naturvardsverke Fack (Sweden), 178
steam, dioxin disposal, 221
Steele, Jesse, 100
steroid metabolism, 55
stillbirths, 41, 85, 161, 177
Stimson, Henry, 148
Stockholm (Sweden), 92, 267
stomach, dioxin poisoning, 35, 104
streptozotoxin, 54
sugar cane, 40, 82, 176
sugar metabolism, disordered, 130
Sullivan, William H., 193
sulphuric acid, 11, 102
sunlight, dioxin degradation, 27, 219
Suskind, Dr. Raymond, 90-1, 100, 161-2
Swanwick, Michael, 110
Sweden, 74, 92, 165, 178
  forestry workers, 52-3, 177-9, 240
  Medical Authorities, 74
  Natural Research Council, 78
  railroad workers, 52
  U.S. Ambassador, 165
Swedish Lapland, 176
Swiss/H/Riop mice strain, 51
Switzerland, 11, 15, 197, 198, 199, 201, 204, 209, 215
  Federal Research Station, 16
Syntex Agribusiness Inc. (U.S.), 126
synthetic acids, 77, 148
synthetic plant hormones, 77, 150

talcum powder, hexachlorophene, infant deaths, 71-3, 74-5, 127
talipes, 176
Tan Son Nhut (South Vietnam), 151
Tanganyika, 150

tanning industry, 35
Tay Ninh City Provincial Hospital (Vietnam), 161
Taylor, Ernest, 113, 115
Taylor, Dr. James, 89
Taylor, General Maxwell, 151
TCDD: *see* dioxin
Telegina, Dr. K.A., 121-2
teratogenic effects: *see* birth defects
termite control, 82
terrorism, Italy, 224
Tesla, Louis, 229
testis, cancer, animals, 51
tests
  animals, 35, 107
  cancer, 51
tetrachlorinated isomers, 19
tetrachlorobenzene, 6, 8, 9, 110
tetrachlorodibenzo-*p*-dioxin: *see* dioxin
tetrachlorofurans, 14
tetrachlorodibenzofurans, 8, 9, 13, 14, 31
tetrachlorophenol, 6, 11
TGWU: *see* Transport & General Workers Union
Thailand, 150
Thalidomide, 216
Thames River (UK), 82
Theofanous, Professor T.G., 223
Thomson, Hayward (US), 132, 172
thymus gland, animals, 32, 46, 57, 107
thyroid cancer, animals, 47
thyroid hormones, 56
timber loss, Vietnam, 166
Times (London), 113
tissue culture tests, 47
Tittabawassee River (U.S.), 15
toluene, 230
tongue, cancer, animals, 47
tonsillitis, 122
tooth defects, 86
Townsend, Dr. D.E., 18
toxic metals, 13
toxic waste disposal, US, 3, 230
trade unions (U.K.), 177, 178
Trades Union Congress (U.K.), 177
Train, Russell, 165
transaminase, 211
Transport & General Workers Union (UK), 116
trichlorodibenzo-*p*-dioxin, 5, 30
trichloroethylene, 230

trichlorophenol, 11, 18, 81-2, 92, 109, 137, 138, 212, 230, 313
  cancer link, soft tissue sarcoma, 52-3
  liver damage, 34
  manufacture, 1, 102-3, 110, 112, 115-6, 123, 127, 131, 198, 207, 209, 223, 224, 239
    accidents, 9, 26, 36, 95ff., 139, 164, 197ff., 237
    simulation, 222
    waste, 122-8
triglyceride, 34, 54, 84, 85, 117, 136
triple dye, bacteriacide, 71
trypsin, 56
Tschirley, Dr. Fred, 156
Tsetse fly, 150
Tuchmann-Duplessis, Dr., 218
Tung, Professor Ton That, 168
Tuyen, Dr. Bach Quoc, 168

Uberti, Francesco, 198
UC-123 aircraft, 188
Ufa Research Institute of Hygiene & Industrial Medicine (USSR), 121
ultraviolet light
  dioxin degradation, 26-7, 115, 128, 221
  skin treatment, 111, 122
Umea (Sweden), 5, 178
United Nations
  Conference on the Human Environment, 1972, 165
  General Assembly, 153, 155
  Security Council, 151, 187
  Vietnam Study, 155
United States Air Force, 147-9, 172, 187, 188, 190
  Epidemiology Division, 169
  herbicide stocks, 162
  Institute of Pathology, 171
  Occupational & Environmental Health Laboratory, 170
  Surgeon General, 170
  Vietnam, 147ff.
United States Army, 132, 148, 149, 161
  Fort Detrick, 148, 149, 150
United States Central Intelligence Agency, 189, 191
United States Congress, 163
United States Embassy (Saigon), 156
United States Government, 132-3, 147, 149, 154, 159-60

United States Government (cont.)
  Department of Agriculture, 156, 160, 162, 164
    Plant Industry Station, 148
  Department of Defense, 148, 150, 154, 155, 159, 162, 163, 172, 189, 191, 193-4
    aerial photographs, Vietnam, 166-7
    Military Assistance Command for Vietnam, 160
    Policy Plans & National Security Council Affairs Office, 159
  Department of Health & Human Services, 232, 240
  Department of Health, Education & Welfare, 126, 158, 160, 230-1, 240
    Center for Disease Control, 123
    see also United States Government, Department of Health & Human Services
  Department of State, 155-6, 157, 160, 162, 165, 188, 189, 191, 193
  Department of the Interior, 160
  Food & Drug Administration, 71, 74, 83, 158-9
  Justice Department, 233
    Supreme Court, 234
    Court of Appeals, 173
  National Security Council, 151, 189
  Science Advisory Committee, 164
United States House of Representatives, Foreign Affairs Committee, 159
United States, National Toxicology Program, 169
United States, Navy, 149
United States, Senate, 163
  Select Committee on Intelligence, 189
  Sub-Committee on the Environment, 126
University College, North Wales (UK), 166
University of California, 74
University of Chicago, 148
University of Cincinnati, 101
University of Hamburg (W. Germany), 89
University of Montana, 152, 188
University of Pavia (Italy), 212
University of Pennsylvania, 134
University of Umea (Sweden), 5
urea nitrogen, 54
urinary epithelium, 47
urinary tract, 36
uroporphyrinogen decarboxylase, 34

Index

uroporphyrins, 129, 130, 200
uroporphyrinuria, 36, 129, 130, 200
  animals, 34
U.S.S.R., 121-2
Utah (US), 25
uterus, cancer, animals, 47

vanadium, 16
Vaterlaus, Dr. Bruno, 199, 201, 203-4
Vatican, 214
Vedinger (W. Germany), 138
vertigo, 99
veterans: see Vietnam veterans
Veterans Administration (US), 170-1, 240
  hospitals, 171
Victoria (Australia), 40
  Minister of Health, Consultative Council (Australia), 176
Victoria, Lake (Kenya), 150
Vientiane (Laos), U.S. Ambassador, 193
Viet Duc hospital (Vietnam), 168
Vietnam, 29, 46, 53, 147ff., 190, 194, 211
  forest areas, 147, 156, 161
  hardwood forests, 161
  herbicide spraying, 1, 2, 3, 8, 40, 127, 132, 137, 147ff., 188, 200, 211, 240
    amounts, 147, 151
    birth defects link, 161-2
    dioxin amounts, 164
    procedure, 151-2, 158
    restriction, 147, 158
    stillbirth link, 161
    suspension, 162
    U.S. scientists' petition, 153
  mangrove forests, 147, 161
  Mekong Delta, 147
  South: see South Vietnam
  United Nations Study, 155
  veterans
    (Australia), 172, 242
      medical investigations, 177
    (US), 2, 29, 132, 169-73, 179, 240, 242
      cancer, 169, 170, 171, 240
      cancer registry, 171
      compensation claims, 132-3, 172-3, 179-80
      medical investigations, 171
      offspring, birth defects, 169, 171, 240
    (Vietnam), 168, 240-1, 242
Vietnamese, birth defects, attitude, 161
vitamin A, 100

vitamin C, 209
vitamin synthesis, 208
viticulture, 148
Volgermeerpolder (Netherlands), 109
vomiting, dioxin poisoning, 36, 99, 188
Von Bettman, Dr. S., 89
Von Zwehl, Herwig, 198, 200
Vos, Dr. J.G., 12
"Vulcanus" incinerator ship, 127-8, 163

Waldvogel, Guy, 199, 203, 204, 224
Walter Reed Army Hospital (US), 171
War Research Service (US), 148
Ward, Dr. Anthony, 117
waste oil, dust supression, 123, 127
waste tips, dioxin disposal, 109, 115
water-based herbicide, 151
water hyacinth, 150
water retention: see edema
Watusi Peninsula (Kenya), 150
weed killers: see herbicides
weight loss, 36, 125
  animals, 35, 84
W. Germany, 46, 53, 89, 91, 105, 106
  Federal Health Office, 177
Westing, Professor Arthur, 152, 160, 162, 164, 165, 188, 189, 190, 191, 194
Wheeler, General Earle G., 190
Whiteside, Thomas, 159, 207
wild mustard, 148
Willard, Paul, 99
Windham College (U.S.), 160
Witt, Professor James, 175
Wolfe, Lieutenant Colonel William, 169, 170
wood preservative, 84, 91
workers
  dioxin exposure, 1, 89, 95ff., 237, 242
    Boehringer (W. Germany), 89, 91-2
    Coalite & Chemical Products Ltd. (UK), 3, 104, 109ff.
    ICMESA (Italy), 200
    Monsanto Chemical Company (US), 90-1
    Philips-Duphar (Netherlands), 108-9
  medical investigations, 33, 52-3, 89, 104-5, 121-2, 129
    Coalite & Chemical Products Ltd., 3, 111, 114, 116-7, 118-21, 139
    Dow Chemical Company (U.S.), 46, 47, 51

workers (*cont.*)
  Monsanto Chemical Company, 84, 90–1, 101–2, 136–7
  Philips-Duphar, 108–9
  mortality studies, 240
    BASF (W. Germany), 53, 104–5
    Dow Chemical Company, 133–4, 178
    Monsanto Chemical Company, 53, 102, 134, 178
    Philips Duphar, 108–9
World Health Organization, 217
World War II, 149, 150

Yale University, 152, 159, 160
Yannacone, Victor, 170–3
Yarham (Australia), 40, 176
Yusho patients, 13, 85–6
  deaths, 85
  medical investigations, 12–13
  symptoms, 85–6

Zack, Judith, 101, 102
Zone A, Seveso, 124, 204, 206, 207, 210, 211, 218, 219, 220–1, 222, 229
Zone B, Seveso, 205, 206, 210, 211, 219, 220, 221
Zone R, Seveso, 205, 219, 220
Zurich (Switzerland), 209